Machine Learning on Geographical Data Using Python

Introduction into Geodata with Applications and Use Cases

Joos Korstanje

Apress®

Machine Learning on Geographical Data Using Python: Introduction into Geodata with Applications and Use Cases

Joos Korstanje
VIELS MAISONS, France

ISBN-13 (pbk): 978-1-4842-8286-1
https://doi.org/10.1007/978-1-4842-8287-8

ISBN-13 (electronic): 978-1-4842-8287-8

Managing Director, Apress Media LLC: Welmoed Spahr
Acquisitions Editor: Celestin Suresh John
Development Editor: Laura Berendson
Coordinating Editor: Mark Powers

Cover designed by eStudioCalamar

Cover image by Shutterstock (www.shutterstock.com)

Distributed to the book trade worldwide by Apress Media, LLC, 1 New York Plaza, New York, NY 10004, U.S.A. Phone 1-800-SPRINGER, fax (201) 348-4505, e-mail orders-ny@springer-sbm.com, or visit www.springeronline.com. Apress Media, LLC is a California LLC and the sole member (owner) is Springer Science + Business Media Finance Inc (SSBM Finance Inc). SSBM Finance Inc is a **Delaware** corporation.

For information on translations, please e-mail booktranslations@springernature.com; for reprint, paperback, or audio rights, please e-mail bookpermissions@springernature.com.

Apress titles may be purchased in bulk for academic, corporate, or promotional use. eBook versions and licenses are also available for most titles. For more information, reference our Print and eBook Bulk Sales web page at http://www.apress.com/bulk-sales.

Any source code or other supplementary material referenced by the author in this book is available to readers on GitHub (https://github.com/Apress). For more detailed information, please visit http://www.apress.com/source-code.

Printed on acid-free paper

Table of Contents

About the Author

 **Joos Korstanje** is a data scientist, with over five years of industry experience in developing machine learning tools. He has a double MSc in Applied Data Science and in Environmental Science and has extensive experience working with geodata use cases. He has worked at a number of large companies in the Netherlands and France, developing machine learning for a variety of tools. His experience in writing and teaching has motivated him to write this book on machine learning for geodata with Python.

About the Technical Reviewer

Xiaochi Liu is a PhD researcher and data scientist at Macquarie University, specializing in machine learning, explainable artificial intelligence, spatial analysis, and their novel application in environmental and public health. He is a programming enthusiast using Python and R to conduct end-to-end data analysis. His current research applies cutting-edge AI technologies to untangle the causal nexus between trace metal contamination and human health to develop evidence-based intervention strategies for mitigating environmental exposure.

Introduction

Spatial data has long been an ignored data type in general data science and statistics courses. Yet at the same time, there is a field of spatial analysis which is strongly developed. Due to differences in tools and approaches, the two fields have long developed in separate environments.

With the popularity of data in many business environments, the importance of treating spatial data is also increasing. The goal of the current book is to bridge the gap between data science and spatial analysis by covering tools of both worlds and showing how to use tools from both to answer use cases.

The book starts with a general introduction to geographical data, including data storage formats, data types, common tools and libraries in Python, and the like. Strong attention is paid to the specificities of spatial data, including coordinate systems and more.

The second part of the book covers a number of methods of the field of spatial analysis. All of this is done in Python. Even though Python is not the most common tool in spatial analysis, the ecosystem has taken large steps in user-friendliness and has great interoperability with machine learning libraries. Python with its rich ecosystem of libraries will be an important tool for spatial analysis in the near future.

The third part of the book covers multiple machine learning use cases on spatial data. In this part of the book, you see that tools from spatial analysis are combined with tools from machine learning and data science to realize more advanced use cases than would be possible in many spatial analysis tools. Specific considerations are needed for applying machine learning to spatial data, due to the specific nature of coordinates and other specific data formats of spatial data.

Source Code

All source code used in the book can be downloaded from `github.com/apress/machine-learning-geographic-data-python`.

PART I

General Introduction

CHAPTER 1

Introduction to Geodata

Mapmaking and analysis of the geographical environment around us have been present in nature and human society for a long time. Human maps are well known to all of us: they are a great way to share information about our environment with others.

Yet communicating geographical instructions is not invented only by the human species. Bees, for example, are well known to communicate on food sources with their fellow hive mates. Bees do not make maps, but, just like us, they use a clearly defined communication system.

As geodata is the topic of this book, I find it interesting to share this out-of-the-box geodata system used by honeybees. Geodata in the bee world has two components: distance and direction.

Honeybee distance metrics

- The round dance: A food source is present less than 50 meters from the hive.

- The sickle dance: Food sources are present between 50 and 150 meters from the hive.

- The waggle (a.k.a. wag-tail) dance: Food sources are over 150 meters from the hive. In addition, the duration of the waggle dance is an indicator of how far over 150 meters the source is located.

Honeybee direction metrics

- Although more complicated, the angle of the dance is known to be an indicator of the angle relative to the sun that bees must follow to get to their food source.

- As the sun changes location throughout the day, bees will update each other by adapting their communication dances accordingly.

© Joos Korstanje 2022
J. Korstanje, *Machine Learning on Geographical Data Using Python*,
https://doi.org/10.1007/978-1-4842-8287-8_1

The human counterpart of geographical communication works a bit better, as we have compasses that point to the magnetic north. Those of you who are familiar with compass use, for example, on boats, may know that even using a compass is not a perfect solution.

The magnetic north changes much less than the position of the sun. What is interesting though is that the magnetic north and the true north are not located at the same exact place. The true north is a fixed location on the globe (the so-called North Pole), but compasses are based on magnetism and therefore point to the magnetic north: a location that moves a little bit every year.

If you are navigating a ship with a compass, you will constantly need to do calculations that convert your magnetic direction measurements into true direction measurements by adding magnetic variation, which is a value that changes depending on where you are on earth.

Reading Guide for This Book

As you will understand from these two examples, working with geodata is a challenge. While identifying locations of points by coordinates may appear simple, the devil really is in the details.

The goal of this book is to go over all those details while working on example code projects in Python. This should give you the fundamental knowledge needed to start working in the interesting domain of geodata while avoiding mistakes. You will then discover numerous ways to represent geodata and learn to work with tools that make working with geodata easier.

After laying the basis, the book will become more and more advanced by focusing on machine learning techniques for the geodata domain. As you may expect, the specificities of the use of geodata make that a lot of standards techniques are not applicable at all, or in other cases, they may need specific adaptations and configurations.

Geodata Definitions

To get started, I want to cover the basics of coordinate systems in the simplest mathematic situation: the Euclidean space. Although the world does not respect the hypothesis made by Euclidean geometry, it is a great entry into the deeper understanding of coordinate systems.

A two-dimensional Euclidean space is often depicted as shown in Figure 1-1.

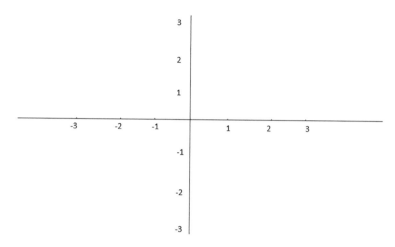

Figure 1-1. *A two-dimensional Euclidean space. Image by author*

Cartesian Coordinates

To locate points in the Euclidean space, we can use the Cartesian coordinate system. This coordinate system specifies each point uniquely by a pair of numerical coordinates. For example, look at the coordinate system in Figure 1-2, in which two points are located: a square and a triangle.

The square is located at x = 2 and y = 1 (horizontal axis). The triangle is located at x = -2 and y = -1.

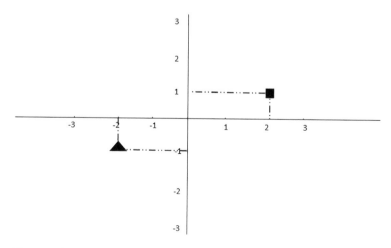

Figure 1-2. *Two points in a coordinate system. Image by author*

The point where the x and y axes meet is called the Origin, and distances are measured from there. Cartesian coordinates are among the most well-known coordinate system and work easily and intuitively in the Euclidean space.

Polar Coordinates and Degrees

A commonly used alternative to Cartesian coordinates is the polar coordinate system. In the polar system, one starts by defining one point as the pole. From this pole starts the polar axis. The graphic in Figure 1-3 shows the idea.

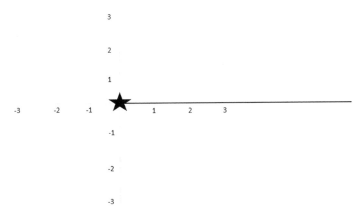

Figure 1-3. *The polar system. Image by author*

In this schematic drawing, the star is designated as the pole, and the thick black line to the right is chosen as the polar axis. This system is quite different from the Cartesian system but still allows us to identify the exact same points: just in a different way.

The points are identified by two components: an angle with respect to the polar axis and a distance. The square that used to be referred to as Cartesian coordinate (2,1) can be referred to by an angle from the polar axis and a distance.

This is shown in Figure 1-4.

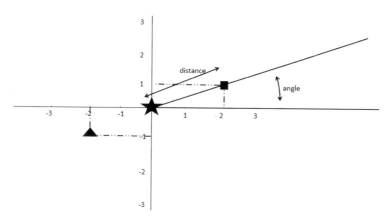

Figure 1-4. *A point in the polar coordinate system. Image by author*

At this point, you can measure the distance and the angle and obtain the coordinate in the polar system. Judged by the eye alone, we could say that the angle is probably more or less 30° and the distance is slightly above 2. We would need to have more precise measurement tools and a more precise drawing for more precision.

There are trigonometric computations that we can use to convert between polar and Cartesian coordinates. The first set of formulas allows you to go from polar to Cartesian:

$$x = r \cos \varphi$$
$$y = r \sin \varphi$$

The letter r signifies the distance and the letter φ is the angle. You can go the other way as well, using the following formulas:

$$r = \sqrt{x^2 + y^2}$$
$$\varphi = \text{atan2}(y, x)$$

As a last part to cover about degrees, I want to mention the equivalence between measuring angles in degrees and in radians. The radian system may seem scary if you are not used to it, but just remember that for every possible angle that you can measure (from 0 to 360) there is a corresponding notation in the radian system. Figure 1-5 shows this.

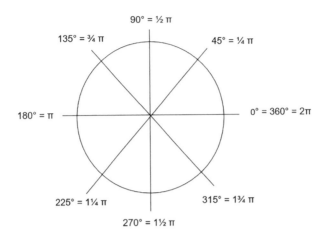

Figure 1-5. *Radians vs. degrees. Image by author*

The Difference with Reality

In reality, you never work with Euclidean space on a map. This is because the world is not flat, but rather a sort of sphere. First of all, it needs to be considered that the object is in three dimensions. More importantly, distances from one point to another need to take into account the specific curvature of the earth at that point. After all, to make it even more difficult, the earth unfortunately is not a perfectly round ball.

In the real world, things have to be much more complicated than in the Euclidean examples. This is done by the Geographic Coordinate System which is adapted to our ellipsoidal reality. In this system, we usually measure a point by a combination of latitude and longitude.

Latitude indicates how high or low on the globe you are with respect to the equator. Longitude tells us how much left or right on the globe you are with respect to the Greenwich meridian.

The earth is split into four quadrants from the zero point at the intersection of the equator and the Greenwich meridian. You have north and south and east and west together making up four quadrants. The North Pole has a latitude of 90 degrees North, and the South Pole is 90 degrees South. Longitude ranges from 180 degrees West to 180 degrees East.

Degrees do not have decimals, but rather can be cut up into minutes. One degree consists of 60 minutes, and one minute consists of 60 seconds.

Geographic Information Systems and Common Tools

As you must understand by now, geodata is an easy way into a headache if you do not have tools that do all the conversion work for you. And we are lucky, as many such tools exist. In this part, let's have a look at a few of the most commonly used tools together with some advantages and disadvantages of each of them.

What Are Geographic Information Systems

GIS or Geographic Information Systems are a special type of database system that is made specifically for geographic data, also called geodata. Those database systems are developed in such a way that problems like coordinate systems and more are not a problem to be solved by the user. It is all done inherently by the system. GIS also stands for the industry that deals with those information systems.

ArcGIS

ArcGIS, made by ESRI, is arguably the most famous software package for working with Geographic Information Systems. It has a very large number of functionalities that can be accessed through a user-friendly click-button system, but visual programming of geodata processing pipelines is also allowed. Python integration is even possible for those who have specific tasks for which there are no preexisting tools in ArcGIS. Among its tools are also AI and data science options.

ArcGIS is a great software for working with geodata. Yet there is one big disadvantage, and that is that it is a paid, proprietary software. It is therefore accessible only to companies or individuals that have no difficulty paying the considerably high price. Even though it may be worth its price, you'll need to be able to pay or convince your company to pay for such software. Unfortunately, this is often not the case.

QGIS and Other Open Source ArcGIS Alternatives

Open source developers have jumped into this open niche of GIS systems by developing open source (and therefore free to use) alternatives. These include QGIS, GRASS GIS, PostGIS, and more.

The clear advantage of this is that they are free to use. Yet their functionality is often much more limited. In most of them, users have the ability to code their own modules in case some of the needed tools are not available.

This approach can be a good fit for your need if you are not afraid to commit to a system like QGIS and fill the gaps that you may eventually encounter.

Python/R Programming

Finally, you can use Python or R programming for working with geodata as well. Programming, especially in Python or R, is a very common skill among data professionals nowadays.

As programming skills were less well spread a few years back, the boom in data science, machine learning, and artificial intelligence has made languages like Python become very commonly spread throughout the workforce.

Now that many are able to code or have access to courses to learn how to code, the need for full software becomes less. The availability of a number of well-functioning geodata packages is enough for many to get started.

Python or R programming is a great tool for treating geodata with common or more modern methods. By using these programming languages, you can easily apply tools from other libraries to your geodata, without having to convert this to QGIS modules, for example.

The only problem that is not very well solved by programming languages is long-term geodata storage. For this, you will need a database. Cloud-based databases are nowadays relatively easy to arrange and manage, and this problem is therefore relatively easily solved.

Standard Formats of Geodata

As you have understood, there are different tools and programming languages that can easily deal with geodata. While doing geodata in Python in this book, we will be generally interested much more in data processing than in data storage.

Yet a full solution for geodata treatment cannot rely on treatment alone. We also need a data format. You will now see a number of common data formats that are used very widely for storing geographical data.

Shapefile

The shapefile is a very commonly used file format for geodata because it is the standard format for ArcGIS. The shapefile is not very friendly for being used outside of ArcGIS, but due to the popularity of ArcGIS, you will likely encounter shapefiles at some point.

The shapefile is not really a single file. It is actually a collection of files that are stored together in one and the same directory, all having the same name. You have the following files that make up a shapefile:

- myfile.shp: The main file, also called the shapefile (confusing but true)

- myfile.shx: The shapefile index file

- myfile.dbf: The shapefile data file that stores attribute data

- myfile.prj: Optional file that stores spatial reference and projection metadata

As an example, let's look at an open data dataset containing the municipalities of the Paris region that is provided by the French government. This dataset is freely available at https://geo.data.gouv.fr/en/datasets/8fadd7040c4b94f2c318a0971e8fae db7b5675d6

On this website, you can download the data in SHP/L93 format, and this will allow you to download a directory with a zip file. Figure 1-6 shows what this contains.

Name	Type	Compress...	Pass...	Size	Ratio	Date modified
Communes_MGP.dbf	OpenOffic...	23 KB	No	117 KB	81%	03/02/2022 20:27
Communes_MGP.prj	PRJ File	1 KB	No	1 KB	38%	03/02/2022 20:27
Communes_MGP.shp	SHP File	919 KB	No	1,439 KB	37%	03/02/2022 20:27
Communes_MGP.shx	SHX File	1 KB	No	2 KB	28%	03/02/2022 20:27

Figure 1-6. *The inside of the shapefile. Image by author*
Data source: Ministry of DINSIC. Original data downloaded from https://geo. data.gouv.fr/en/datasets/8fadd7040c4b94f2c318a0971e8faedb7b5675d6, updated on 1 July 2016. Open Licence 2.0 (www.etalab.gouv.fr/wp-content/ uploads/2018/11/open-licence.pdf)

As you can see, there are the .shp file (the main file), the .shx file (the index file), the .dbf file containing the attributes, and finally the optional .prj file.

For this exercise, if you want to follow along, you can use your local environment or a Google Colab Notebook at https://colab.research.google.com/.

You have to make sure that in your environment, you install geopandas:

```
!pip install geopandas
```

Then, make sure that in your environment you have a directory called Communes_MGP.shp in which you have the four files:

- Communes_MGP.shp

- Communes_MGP.dbf

- Communes_MGP.prj

- Communes_MGP.shx

In a local environment, you need to put the "sample_data" file in the same directory as the notebook, but when you are working on Colab, you will need to upload the whole folder to your working environment, by clicking the folder icon and then dragging and dropping the whole folder onto there. You can then execute the Python code in Code Block 1-1 to have a peek inside the data.

Code Block 1-1. Importing the shapefile

```
import geopandas as gpd
shapefile = gpd.read_file("sample_data/Communes_MGP.shp")
print(shapefile)
```

You'll see the result in Figure 1-7.

```
     ID_APUR   ...                                          geometry
0       1295   ...   POLYGON ((643143.540 6874983.000, 643145.170 6...
1       1296   ...   POLYGON ((650150.470 6868653.663, 650149.336 6...
2       1297   ...   POLYGON ((648778.017 6874199.664, 648780.177 6...
3        543   ...   POLYGON ((657415.714 6873254.944, 657415.592 6...
4        544   ...   POLYGON ((653644.775 6849095.507, 653647.315 6...
..       ...   ...                                               ...
147      885   ...   POLYGON ((653492.987 6865245.556, 653409.972 6...
148      886   ...   POLYGON ((655238.908 6866974.758, 655290.661 6...
149      887   ...   POLYGON ((656928.376 6864099.214, 656933.359 6...
150      888   ...   POLYGON ((646773.335 6864505.116, 647009.256 6...
151      889   ...   POLYGON ((659380.730 6860752.310, 659373.929 6...

[152 rows x 37 columns]
```

Figure 1-7. *The data in Python. Image by author*
Data source: Ministry of DINSIC. Original data downloaded from https://geo.
data.gouv.fr/en/datasets/8fadd7040c4b94f2c318a0971e8faedb7b5675d6,
updated on 1 July 2016. Open Licence 2.0 (www.etalab.gouv.fr/wp-content/
uploads/2018/11/open-licence.pdf)

To make something more visual, you can use the code in Code Block 1-2.

Code Block 1-2. Plotting the shapefile

```
shapefile.plot()
```

You will obtain the map corresponding to this dataset as in Figure 1-8.

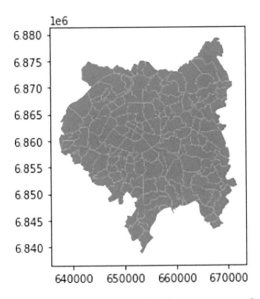

Figure 1-8. *The map resulting from Code Block 1-2. Image by author*
Data source: Ministry of DINSIC. Original data downloaded from https://geo.
data.gouv.fr/en/datasets/8fadd7040c4b94f2c318a0971e8faedb7b5675d6,
*updated on 1 July 2016. Open Licence 2.0 (*www.etalab.gouv.fr/wp-content/
uploads/2018/11/open-licence.pdf*)*

Google KML File

You are probably familiar with Google Earth: one of the dominating map-based applications of our time. Google has popularized the KML file for geodata. It is an XML-based text file that can contain geometry data.

The .KMZ file is a compressed version of the KML file. You can decompress it using any unzipping tool and obtain a KML file.

As an example, let's look at the exact same database as before, which is located at France's open geodata platform:

Ministry of DINSIC. Original data downloaded from https://geo.data.gouv.fr/en/
datasets/8fadd7040c4b94f2c318a0971e8faedb7b5675d6, updated on 1 July 2016. Open
Licence 2.0 (www.etalab.gouv.fr/wp-content/uploads/2018/11/open-licence.pdf)

In the resources part, you'll see that this map of the Paris region's municipalities is also available in the KML format. Download it and you'll obtain a file called Communes_MGP.kml.

If you try opening the file with a text editor, you'll find that it is an XML file (very summarized, XML is a data storage pattern that can be recognized by many < and > signs).

Compared to the shapefile, you can see that KML is much easier to understand and to parse. A part of the file contents is shown in Figure 1-9.

```
Communes_MGP.kml - Notepad

File  Edit  Format  View  Help
<?xml version="1.0" encoding="utf-8" ?>
<kml xmlns="http://www.opengis.net/kml/2.2">
<Document id="root_doc">
<Schema name="Communes_MGP" id="Communes_MGP">
        <SimpleField name="ID_APUR" type="float"></SimpleField>
        <SimpleField name="N_SQ_CAB" type="float"></SimpleField>
        <SimpleField name="C_INSEE" type="float"></SimpleField>
        <SimpleField name="L_CAB" type="string"></SimpleField>
        <SimpleField name="B_BOIS" type="string"></SimpleField>
        <SimpleField name="L_BOIS" type="string"></SimpleField>
        <SimpleField name="N_SQ_AR" type="string"></SimpleField>
        <SimpleField name="N_SQ_CO" type="float"></SimpleField>
        <SimpleField name="LIEN_INSEE" type="string"></SimpleField>
        <SimpleField name="POP_08" type="float"></SimpleField>
        <SimpleField name="POP_13" type="float"></SimpleField>
        <SimpleField name="POP_14" type="float"></SimpleField>
        <SimpleField name="08_EVO_13" type="float"></SimpleField>
        <SimpleField name="08_PEVO_13" type="int"></SimpleField>
        <SimpleField name="08_EVO_14" type="float"></SimpleField>
        <SimpleField name="08_PEVO_14" type="int"></SimpleField>
        <SimpleField name="13_EVO_14" type="float"></SimpleField>
        <SimpleField name="13_PEVO_14" type="float"></SimpleField>
        <SimpleField name="QPV_13" type="float"></SimpleField>
        <SimpleField name="PROP_QPV" type="float"></SimpleField>
        <SimpleField name="POP_15_19a" type="float"></SimpleField>
        <SimpleField name="POP_20_24a" type="float"></SimpleField>
        <SimpleField name="POP_25_39a" type="float"></SimpleField>
        <SimpleField name="POP_40_54a" type="float"></SimpleField>
        <SimpleField name="POP_55_64a" type="float"></SimpleField>
        <SimpleField name="POP_65_79a" type="float"></SimpleField>
        <SimpleField name="POP_sup80a" type="float"></SimpleField>
        <SimpleField name="PROP_15_19" type="float"></SimpleField>
        <SimpleField name="PROP_20_24" type="float"></SimpleField>
        <SimpleField name="PROP_25_39" type="float"></SimpleField>
        <SimpleField name="PROP_40_54" type="float"></SimpleField>
        <SimpleField name="PROP_55_64" type="float"></SimpleField>
        <SimpleField name="PROP_65_79" type="float"></SimpleField>
        <SimpleField name="PROP_sup80" type="float"></SimpleField>
        <SimpleField name="SHAPE_Leng" type="float"></SimpleField>
        <SimpleField name="SHAPE_Area" type="float"></SimpleField>
</Schema>
<Folder><name>Communes_MGP</name>
```

Figure 1-9. *The KML file content. Image by author*
Data source: Ministry of DINSIC. Original data downloaded from https://geo.data.gouv.fr/en/datasets/8fadd7040c4b94f2c318a0971e8faedb7b5675d6, updated on 1 July 2016. Open Licence 2.0 (www.etalab.gouv.fr/wp-content/uploads/2018/11/open-licence.pdf)

To get a KML file into Python, we can again use geopandas. This time, however, it is a bit less straightforward. You'll also need the Fiona package to obtain a KML driver. The total code is shown in Code Block 1-3.

Code Block 1-3. Importing the KML file

```
import fiona
gpd.io.file.fiona.drvsupport.supported_drivers['KML'] = 'rw'

import geopandas as gpd
kmlfile = gpd.read_file("Communes_MGP.kml")
print(kmlfile)
```

You'll then see the exact same geodataframe as before, which is shown in Figure 1-10.

```
    Name Description                                        geometry
0                      POLYGON ((2.22337 48.97227, 2.22340 48.97224, ...
1                      POLYGON ((2.31982 48.91594, 2.31981 48.91594, ...
2                      POLYGON ((2.30043 48.96570, 2.30046 48.96569, ...
3                      POLYGON ((2.41849 48.95784, 2.41849 48.95765, ...
4                      POLYGON ((2.36963 48.74031, 2.36968 48.73948, ...
..     ...          ...                                             ...
147                    POLYGON ((2.36580 48.88554, 2.36469 48.88437, ...
148                    POLYGON ((2.38943 48.90122, 2.39014 48.90108, ...
149                    POLYGON ((2.41277 48.87547, 2.41284 48.87524, ...
150                    POLYGON ((2.27427 48.87837, 2.27749 48.87796, ...
151                    POLYGON ((2.44652 48.84554, 2.44643 48.84504, ...

[152 rows x 3 columns]
```

Figure 1-10. *The KML data shown in Python. Image by author*
Data source: Ministry of DINSIC. Original data downloaded from https://geo. data.gouv.fr/en/datasets/8fadd7040c4b94f2c318a0971e8faedb7b5675d6, updated on 1 July 2016. Open Licence 2.0 (www.etalab.gouv.fr/wp-content/ uploads/2018/11/open-licence.pdf)

As before, you can plot this geodataframe to obtain a basic map containing the municipalities of the area of Paris and around. This is done in Code Block 1-4.

Code Block 1-4. Plotting the KML file data

```
kmlfile.plot()
```

The result is shown in Figure 1-11.

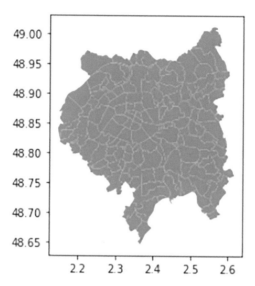

Figure 1-11. *The plot resulting from Code Block 1-4. Screenshot by author*
Data source: Ministry of DINSIC. Original data downloaded from https://geo.
data.gouv.fr/en/datasets/8fadd7040c4b94f2c318a0971e8faedb7b5675d6,
updated on 1 July 2016. Open Licence 2.0 (www.etalab.gouv.fr/wp-content/
uploads/2018/11/open-licence.pdf)

An interesting point here is that the coordinates do not correspond with the map that was generated from the shapefile. If you've read the first part of this chapter, you may have a hinge on how this is caused by coordinate systems. We'll get into this in much more detail in Chapter 2.

GeoJSON

The json format is a data format that is well known and loved by developers. Json is much used in communication between different information systems, for example, in website and Internet communication.

The json format is loved because it is very easy to parse, and this makes it a perfect storage for open source and other developer-oriented tools.

Json is a key-value dataset, which is much like the dictionary in Python. The whole is surrounded by accolades. As an example, I could write myself as a json object as in this example:

```
{   'first_name': 'joos',
    'last_name': 'korstanje',
    'job': 'data scientist' }
```

As you can see, this is a very flexible format, and it is very easy to adapt to all kinds of circumstances. You might easily add GPS coordinates like this:

```
{   'first_name': 'joos',
    'last_name': 'korstanje',
    'job': 'data scientist',
    'latitude': '48.8566° N',
    'longitude': '2.3522° E' }
```

GeoJSON is a json-based format that defines a specific, standardized way to deal with storing coordinates (not just points but also lines and polygons) in the json format.

The Paris municipalities map that you've downloaded before is also available in the geojson format. Download it over here (click GeoJSON in the Resources part):

```
https://geo.data.gouv.fr/en/datasets/8fadd7040c4b94f2c318a0971e8fae
db7b5675d6
```

You will obtain a file called Communes_MGP.json. When opening it with notepad or any other text editor, you'll see that it is a json format (shown in Figure 1-12). Of course, it is the exact same data: only the storage format changes.

Communes_MGP.json - Notepad

File Edit Format View Help

{
"type": "FeatureCollection",
"name": "Communes_MGP",
"crs": { "type": "name", "properties": { "name": "urn:ogc:def:crs:OGC:1.3:CRS84" } },
"features": [
{ "type": "Feature", "properties": { "ID_APUR": 1295, "N_SQ_CAB": 950001524, "C_INSEE": 95018, "L_CAB": "Argenteuil", "B_BOIS": "Non", "L_BOIS": null, "N_SQ_AR": null, "N_SQ_CO": 950001524, "LIEN_INSEE":
"http:\/\/www.insee.fr\/fr\/themes\/tableau_local.asp?ref_id=POP&millesime=2013&typgeo=COM&search=95018", "POP_08": 105256, "POP_13": 108414, "POP_14": 108865, "08_EVO_13": 3158, "08_PEVO_13": 3,
"08_EVO_14": 3609, "08_PEVO_14": 3, "13_EVO_14": 451, "13_PEVO_14": 0, "QPV_13": null, "PROP_QPV": null, "POP_15_19a": 6830, "POP_20_24a": 7299, "POP_25_39a": 23765, "POP_40_54a": 21154, "POP_55_64a":
10919, "POP_65_79a": 8507, "POP_sup80a": 3245, "PROP_15_19": 6.3, "PROP_20_24": 6.73, "PROP_25_39": 21.92, "PROP_40_54": 19.51, "PROP_55_64": 10.07, "PROP_65_79": 7.85, "PROP_sup80": 2.99, "SHAPE_Leng":
20392.54504, "SHAPE_Area": 17384988.794580001 }, "geometry": { "type": "Polygon", "coordinates": [[[2.2337490140284141, 48.972274499807163], [2.22339769407591242, 48.972239216199149], [
2.223419779336183, 48.972324430025878], [2.22346118581813, 48.972359859523031], [2.22350313221406, 48.97235100901914], [2.223886939511442, 48.972269990590284], [2.22407375237242,
48.972026508235956], [2.224363219308333, 48.971842321793901], [2.224659007443718, 48.971591359967900], [2.22473469484987, 48.971526382252705], [2.22517753377218, 48.971242736682328], [
2.22546527924582, 48.971127690207332], [2.225754020340103, 48.971026038576142], [2.225755645264863, 48.971020033965345], [2.22580743775069, 48.971010054574734], [2.226196380816392,
48.970935108689012], [2.226745929397998, 48.970829212785731], [2.226853695072531, 48.970800032776], [2.226966495373921, 48.970785861781742], [2.227157324077944, 48.970748412206895], [
2.230366883855913, 48.970130030040551], [2.230350822685118, 48.970024712397404], [2.230344965260205, 48.969986725617254], [2.230294422769126, 48.969700972935833], [2.23027545606258,
48.969651039733371], [2.23025117040491, 48.969587117508908], [2.230240225154185, 48.969567274106872], [2.230231256785682, 48.969551025183172], [2.230303012509, 48.969517487001632], [
2.23034900066078, 48.969483519928268], [2.23039503623355, 48.969446315770249], [2.23067793355917, 48.969205147896368], [2.230691412607049, 48.969198849537669], [2.23098082963364,
2.231317728220543, 48.968567350762243], [2.231385655071839, 48.968472103669882], [2.231414200107715, 48.968416801577717], [2.231455522388883, 48.96831949546388], [2.231585295507175,
2.968004323060654], [2.231619357778453, 48.967899146677468], [2.231614788853466, 48.967876697664671], [2.231611972103185, 48.967862858712564], [2.231618147794361, 48.967842394887455], [
2.231684330701387, 48.967670968082402], [2.23174385270237, 48.967506602794693], [2.231752068917827, 48.967486691591162], [2.232019553201582, 48.967566551493692], [2.23225653477707,
2.234000769029933, 48.968073821413171], [2.234511038291322, 48.968186520742591], [2.234606656853103, 48.968204557687631], [2.234732495559732, 48.968220307811563], [2.23555813111851,
48.968324942237013], [2.235783710325329, 48.968319995556854], [2.236079850239328, 48.968300200982194], [2.236419729265671, 48.968250394422264], [2.236566668325624, 48.968215181192157], [
2.237616443035574, 48.967928966618629], [2.237911240936979, 48.967778953221533], [2.238049980057923, 48.967723180881372], [2.233881948121043, 48.967628118002686], [2.238443811250435,
48.967607698023137], [2.238699080021539, 48.967496646021015], [2.238945051465508, 48.967444137050], [2.23911943350559, 48.967407042500874], [2.239904358759927, 48.967250330921217], [
2.240186525581735, 48.967195446863359], [2.240446339809487, 48.967165366388755], [2.240656051394752, 48.967069488246658], [2.240796094191257, 48.966999687086087], [2.240912068264596,
48.96693324564076], [2.240968908136366, 48.966903749590905], [2.241020191131376, 48.966887390873403], [2.241045478582525, 48.966879324372599], [2.241084585232042, 48.966876334474996], [
2.241491070377024, 48.966983636049459], [2.241500095895465, 48.966902883789238], [2.241838859769496, 48.966834882724349], [2.241973107738695, 48.966826468853029], [2.242047712083794,
48.966815070003982], [2.242362589161959, 48.966745927293452], [2.242445584105407, 48.966721092134286], [2.242778611227746, 48.966578296646205], [2.242884419626753, 48.966606453020098], [
2.242985528504255, 48.966803457672313], [2.243022091913099, 48.966843795963001], [2.243025460369443, 48.966846964673863], [2.243485849388345, 48.966731814266982], [2.243521594957071,
48.967233567092705], [2.243661405897487, 48.967133878371577], [2.24365731298808, 48.967113124666982], [2.243724683237948, 48.967087614610804], [2.243784673840892, 48.967066771009575], [
2.243967630166656, 48.967024398909388], [2.244212709292301, 48.966983588000161], [2.244451395325683, 48.966959732535315], [2.244526337165285, 48.966953280264505], [2.244549017030872,
48.966896409353438], [2.244798528077004, 48.966851308665802], [2.244834008003739, 48.966847306077405], [2.244506489925842, 48.966833774291928], [2.240594689925842, 48.966807556079722], [
2.245265391513577, 48.966774662202461], [2.245336198483149, 48.966761348944956], [2.245369419539166, 48.966778014650608], [2.245398601854132, 48.966771813450841], [2.245636684822717,
48.966784696901281], [2.245728448004949, 48.966695193336186], [2.245916491580987, 48.966570280934], [2.245907089719464, 48.966650282990059], [2.245886258841237, 48.966589242572702], [
2.246109467331524, 48.966527028880549], [2.246614212399769, 48.966384226529136], [2.246660017049299, 48.966371384910602], [2.246840744374147, 48.966325754354472], [2.246861256139298,
48.966320581068118], [2.246917190290268, 48.966306184276547], [2.247080686635018, 48.966230233236587], [2.247223706049243, 48.966166296297247], [2.247336321838491, 48.966120238334661], [
2.247345661932439, 48.966116609941295], [2.247328723663816, 48.966097888717783], [2.247389795122106, 48.966077676868376], [2.247423474689507, 48.966067627570141], [2.247438491995735,
48.966063146302169], [2.247610230126931, 48.966021598419786], [2.247636147103298, 48.966014476917834], [2.24764189075019, 48.966023415738128], [2.247680634871555, 48.966007651765509], [

Figure 1-12. *The content in json format. Image by author*
Data source: Ministry of DINSIC. Original data downloaded from https://geo.
data.gouv.fr/en/datasets/8fadd7040c4b94f2c318a0971e8faedb7b5675d6,
updated on 1 July 2016. Open Licence 2.0 (www.etalab.gouv.fr/wp-content/
uploads/2018/11/open-licence.pdf)

You can get a GeoJSON file easily into the geopandas library using the code in Code Block 1-5.

Code Block 1-5. Importing the geojson

```
import geopandas as gpd
geojsonfile = gpd.read_file("Communes_MGP.json")
print(geojsonfile)
```

As expected, the data looks exactly like before (Figure 1-13). This is because it is transformed into a geodataframe, and therefore the original representation as json is not maintained anymore.

```
        ID_APUR  ...                                               geometry
0           1295  ...  POLYGON ((2.22337 48.97227, 2.22340 48.97224, ...
1           1296  ...  POLYGON ((2.31982 48.91594, 2.31981 48.91594, ...
2           1297  ...  POLYGON ((2.30043 48.96570, 2.30046 48.96569, ...
3            543  ...  POLYGON ((2.41849 48.95784, 2.41849 48.95765, ...
4            544  ...  POLYGON ((2.36963 48.74031, 2.36968 48.73948, ...
..           ...  ...                                               ...
147          885  ...  POLYGON ((2.36580 48.88554, 2.36469 48.88437, ...
148          886  ...  POLYGON ((2.38943 48.90122, 2.39014 48.90108, ...
149          887  ...  POLYGON ((2.41277 48.87547, 2.41284 48.87524, ...
150          888  ...  POLYGON ((2.27427 48.87837, 2.27749 48.87796, ...
151          889  ...  POLYGON ((2.44652 48.84554, 2.44643 48.84504, ...

[152 rows x 37 columns]
```

Figure 1-13. *The geojson content in Python. Image by author*
Data source: Ministry of DINSIC. Original data downloaded from $https://geo.$
$data.gouv.fr/en/datasets/8fadd7040c4b94f2c318a0971e8faedb7b5675d6,$
updated on 1 July 2016. Open Licence 2.0 ($www.etalab.gouv.fr/wp-content/$
$uploads/2018/11/open-licence.pdf$)

You can make the plot of this geodataframe to obtain a map, using the code in Code Block 1-6.

Code Block 1-6. Plotting the geojson data

```
geojsonfile.plot()
```

The resulting plot is shown in Figure 1-14.

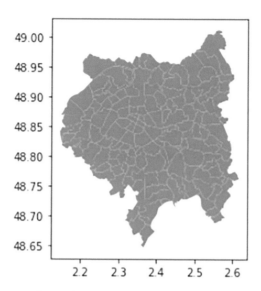

Figure 1-14. *The plot resulting from Code Block 1-6. Image by author
Data source: Ministry of DINSIC. Original data downloaded from* https://geo.
data.gouv.fr/en/datasets/8fadd7040c4b94f2c318a0971e8faedb7b5675d6,
*updated on 1 July 2016. Open Licence 2.0 (*www.etalab.gouv.fr/wp-content/
uploads/2018/11/open-licence.pdf*)*

TIFF/JPEG/PNG

Image file types can also be used to store geodata. After all, many maps are 2D images
that lend themselves well to be stored as an image. Some of the standard formats to store
images are TIFF, JPEG, and PNG.

- The TIFF format is an uncompressed image. A georeferenced TIFF
  image is called a GeoTIFF, and it consists of a directory with a TIFF
  file and a tfw (world file).

- The better-known JPEG file type stores compressed image data.
  When storing a JPEG in the same folder as a JPW (world file), it
  becomes a GeoJPEG.

- The PNG format is another well-known image file format. You can
  make this file into a GeoJPEG as well when using it together with a
  PWG (world file).

Image file types are generally used to store raster data. For now, consider that raster data is image-like (one value per pixel), whereas vector data contains objects like lines, points, and polygons. We'll get to the differences between raster and vector data in a next chapter.

On the following website, you can download a GeoTIFF file that contains an interpolated terrain model of Kerbernez in France:

https://geo.data.gouv.fr/en/datasets/b0a420b9e003d45aaf0670446f0d600df 14430cb

You can use the code in Code Block 1-7 to read and show the raster file in Python.

Code Block 1-7. Read and show the raster data

```
pip install rasterio
import rasterio
from rasterio.plot import import show
fp = r'ore-kbz-mnt-litto3d-5m.tif'
img = rasterio.open(fp)
show(img)
```

Note   Depending on your OS, you may obtain a .tiff file format rather than a .tif when downloading the data. In this case, you can simply change the path to become .tiff, and the result should be the same. In both cases, you will obtain the image shown in Figure 1-15.

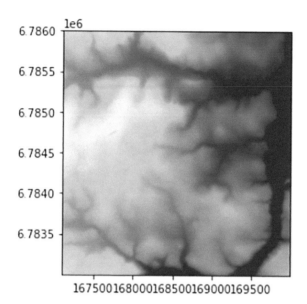

Figure 1-15. *The plot resulting from Code Block 1-7. Image by author Data source: Ministry of DINSIC. Original data downloaded from* `https://geo. data.gouv.fr/en/datasets/b0a420b9e003d45aaf0670446f0d600df14430cb,` *updated on "unknown." Open Licence 2.0 (*`www.etalab.gouv.fr/wp-content/ uploads/2018/11/open-licence.pdf`*)*

It is interesting to look at the coordinates and observe that this file's coordinate values are relatively close to the first file.

CSV/TXT/Excel

The same file as used in the first three examples is also available in CSV. When downloading it and opening it with a text viewer, you will observe something like Figure 1-16.

Figure 1-16. *The contents of the CSV file. Image by author*
Data source: Ministry of DINSIC. Original data downloaded from `https://geo.`
`data.gouv.fr/en/datasets/b0a420b9e003d45aaf0670446f0d600df14430cb`,
updated on "unknown." Open Licence 2.0 (`www.etalab.gouv.fr/wp-content/`
`uploads/2018/11/open-licence.pdf`*)*

The important thing to take away from this part of the chapter is that geodata is "just data," but with geographic references. These can be stored in different formats or in different coordinate systems to make things complicated, but in the end you must simply make sure that you have some sort of understanding of what you have in your data.

You can use many different tools for working with geodata. The goal of those tools is generally to make your life easier. As a last step for this introduction, let's have a short introduction to the different Python tools that you may encounter on your geodata journey.

Overview of Python Tools for Geodata

Here is a list of Python packages that you may want to look into on your journey into geodata with Python:

Geopandas

General GIS tool with a pandas-like code syntax that makes it very accessible for the data science world.

Fiona

Reading and writing geospatial data.

Rasterio

Python package for reading and writing raster data.

GDAL/OGR

A Python package that can be used for translating between different GIS file formats.

RSGISLIB

A package containing remote sensing tools together with raster processing and analysis.

PyProj

A package that can transform coordinates with multiple geographic reference systems.

Geopy

Find postal addresses using coordinates or the inverse.

Shapely

Manipulation of planar geometric objects.

PySAL

Spatial analysis package in Python.

Scipy.spatial

Spatial algorithms based on the famous scipy package for data science.

Cartopy

Package for drawing maps.

GeoViews

Package for interactive maps.

A small reminder: As Python is an open source environment and those libraries are mainly developed and maintained by unpaid open source developers, there is always that chance that something changes or becomes unavailable. This is the risk of working with open source. In most cases, there are no such big problems, but they can and do sometimes happen.

Key Takeaways

1. Cartesian coordinates and polar coordinates are two alternative coordinate systems that can indicate points in a two-dimensional Euclidean space.

2. The world is an ellipsoid, which makes the two-dimensional Euclidean space a bad representation. Other coordinate systems exist for this real-world scenario.

3. Geodata is data that contains geospatial references. Geodata can come in many different shapes and sizes. As long as you have software implementation (or the skills to build it), you will be able to convert between data formats.

4. A number of Python packages exist that do a lot of the heavy lifting for you.

5. The advantage of using Python is that you can have a lot of autonomy on your geodata treatment and that you can benefit from the large number of geodata and other data science and AI packages in the ecosystem.

6. A potential disadvantage of Python is that the software is open source, meaning that you have no guarantee that your preferred libraries still exist in the future. Python is also not suitable for long-term data storage and needs to be complemented with such a data storage solution (e.g., databases or file storage).

CHAPTER 2

Coordinate Systems and Projections

In the previous chapter, you have seen an introduction to coordinate systems. You saw an example of how you can use Cartesian coordinates as well as polar coordinates to identify points on a flat, two-dimensional Euclidean space. It was already mentioned at that point that the real-world scenario is much more complex.

When you are making maps, you are showing things (objects, images, etc.) that are located on earth. Earth does not respect the rules that were shown in the Euclidean example because Earth is an ellipsoid: a ball form that is not perfectly round. This makes map and coordinate system calculations much more complex than what high-school mathematics teaches us about coordinates.

To make the problem clearer, let's look at an example of airplane navigation. Airplane flights are a great example to illustrate the problem, as they generally cover long distances. Taking into account the curvature of the earth really doesn't matter much when measuring the size of your terrace, but it does make a big impact when moving across continents.

Imagine you are flying from Paris to New York using this basic sketch of the world's geography. You are probably well aware of such an organization of the world's map on a two-dimensional image.

A logical first impression would be that to go from Madrid to New York in the quickest way, we should follow a line parallel from the latitude lines. Yet (maybe surprisingly at first) this is **not** the shortest path. An airplane would better curve via the north!

The reason for this is that the more you move to the north, the shorter the latitude lines actually are. Latitude lines go around the earth, so at the North Pole you have a length of zero, and at the equator, the middle line is the longest possible. The closer to the poles, the shorter the distance to go around the earth.

© Joos Korstanje 2022
J. Korstanje, *Machine Learning on Geographical Data Using Python*,
https://doi.org/10.1007/978-1-4842-8287-8_2

As this example takes place in the northern hemisphere, the closest pole is the North Pole. By curving north on the northern hemisphere (toward the pole), an airplane can get to its destination with fewer kilometers. Figure 2-1 illustrates this.

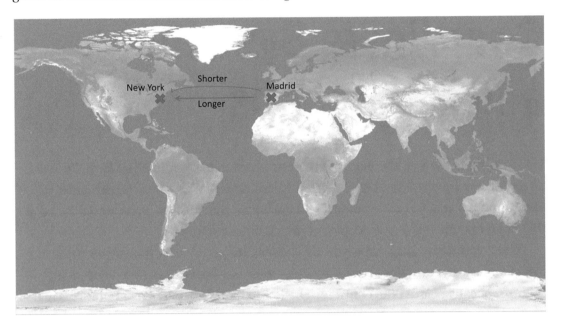

Figure 2-1. *Airplane routes are not straight on a map*
Image adapted from https://en.wikipedia.org/wiki/World_map#/media/ File:Blue_Marble_2002.png *(Public Domain Image. 10 February 2002)*

Let's now consider an example where you are holding a round soccer ball. When going from one point to another on a ball, you will intuitively be able to say which path is the fastest. If you are looking straight at the ball, when following your finger going from one point to another, you will see your hand making a shape like in Figure 2-2.

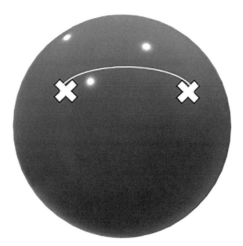

Figure 2-2. *The shortest path on a ball is not a straight line in two-dimensional view. Image by author*

When making maps, we cannot plot in three dimensions, and we, therefore, need to find some way or another to put a three-dimensional path onto a two-dimensional image.

Many map makers have proposed all sorts of ways to solve this unsolvable problem, and the goal of this chapter is to help you understand how to deal effectively with those 3D to 2D mapping distortions that will be continuously looking to complexify your work on geodata.

Coordinate Systems

While the former discussion was merely intuitive, it is now time to slowly get to more official definitions of the concepts that you have seen. As we are ignoring the height of a point (e.g., with respect to sea level) for the moment, we can identify three types of coordinate systems:

- Geographic Coordinate Systems

- Projected Coordinate Systems

- Local Coordinate Systems

Let's go over all three of them.

Geographic Coordinate Systems

Geographic Coordinate Systems are the coordinate systems that we have been talking about in the previous part. They respect the fact that the world is an ellipsoid, and they, therefore, express points using degrees or radians latitude and longitude.

As they respect the ellipsoid property of the earth, it is very hard to make maps or plots with such coordinate systems.

Latitude and Longitude

In Geographic Coordinate Systems, we speak of latitude and longitude. Generally, the degrees are either expressed together with a mention of North (above the equator) and South (below the equator), East (east from Greenwich meridian), or West (west from Greenwich meridian). East/West and North/South can also be written as negative and positive coordinates. East and North take positive, whereas South and West take negative.

Although this is standard practice, this can change depending on the exact definition of the Geographic Coordinate System you are using. In theory, anyone could invent a new coordinate system and make it totally different. In practice, there are some standards that everyone is used to, and therefore it is best to use those.

WGS 1984 Geographic Coordinate System

The WGS 1984, also called WGS 84 or EPSG:4326, is one of the most used Geographic Coordinate Systems. It is also the reference coordinate system of GPS (Global Positioning System) which is used in very many applications. Let's dive into how the WGS 84 was designed.

The WGS 1984 was designed with the goal to have a coordinate origin located at the center of mass of the Earth. The reference meridian or zero meridian is the IERS Reference Meridian. It is very close to the Greenwich meridian: only 5.3 arc seconds or 102 meters to the east.

There are more specific definitions that define the WGS 84, yet at this point, the information becomes very technical. To quote from the Wikipedia page of the WGS 84:

The WGS 84 datum surface is an oblate spheroid with equatorial radius a = 6378137 m at the equator and flattening f = 1/298.257223563. The refined value of the WGS 84 gravitational constant (mass of Earth's atmosphere included) is GM = 3986004.418×108 m3/s2. The angular velocity of the Earth is defined to be ω = 72.92115×10−6 rad/s.

You are absolutely not required to memorize any of those details. I do hope that it gives you an insight into how detailed a definition of Geographic Coordinate Systems has to be. This explains how it is possible that other people and organizations have identified alternate definitions. This is why there are many coordinate systems out there and also one of the reasons why working with geospatial data can be hard to grasp in the beginning.

Other Geographic Coordinate Systems

As an example, ETRS89 is a coordinate system that is recommended by the European Union for use in geodata in Europe. It is very close to the WGS 84 but has some minor differences that make the ETRS89 not subject to change due to continental drift. As you will understand, continental drift is very slow, and taking things like this into account is partly based on theory rather than practical importance.

Another example is the NAD-83 system, which is used mainly for North America. As the Earth is an imperfect ellipsoid, the makers of this system wanted to take into account how much North America deviates from an ellipsoid to perform better in North America.

Projected Coordinate Systems

When making maps, we need to convert the three-dimensional earth into two-dimensional form. This task is impossible to do perfectly: imagining taking a round plastic soccer ball and cutting it open to make it flat. You will never be able to paste this round ball into a flat rectangle without going through an immense struggle to find a way to cut the stuff so that it looks more or less coherent.

Luckily, many great minds have come before us to take this challenge and find great ways to "cut up the ellipsoid earth and paste it on a rectangle." We call this projecting, as you could imagine that the 3D geographic coordinates get matched with a location on the square. This projects everything in between to its corresponding location.

X and Y Coordinates

When working with Projected Coordinate Systems, we do not talk about latitude and longitude anymore. As latitude and longitude are relevant only for measurements on the globe (ellipsoid), but on a flat surface, we can drop this complexity. Once the three-dimensional lat/long coordinates have been converted to the coordinates of their projection, we simply talk about x and y coordinates.

X is generally the distance to the east starting from the origin and y the distance to the north starting from the origin. The location of the origin depends on the projection that you are using. The measurement unit also changes from one Projected Coordinate System to another.

Four Types of Projected Coordinate Systems

As you have understood, there will be a distortion in all Projected Coordinate Systems. As the projections place locations on a three-dimensional ball-like form onto a flat two-dimensional rectangle, some features of reality can be maintained as others get lost.

There is no projection better than the other. The question here is rather which types of features you want to maintain and you are accepting to lose. There are four categories of Projected Coordinate Systems, all making sure that one aspect of reality is perfectly maintained on the projection.

When choosing a map projection for your project, the key choice is to decide which distortions will be the most appropriate and which features you absolutely want to maintain.

Equal Area Projections

The first type of projection is one that preserves the area of specific features. You may not be aware of it, but many maps that you see do not respect equal area. As an example, if you zoom out a bit on Google Maps, you will see that Greenland is huge in the projection that was chosen by Google Maps.

For some use cases, this can be a big problem. Equal area projections are there to make sure that the surface area of specific features stays the same (e.g., countries or others). The cost of this is that other features like shapes and angles may become distorted. You may end up with a projection in which Greenland respects its real area, but the shape is not perfectly represented as a result.

Example 1: Mollweide Projection

The Mollweide projection is an example of an equal area projection. It is also known as the Babinet projection, the homolographic projection, and the elliptical projection. The world looks like Figure 2-3 when projected using Mollweide.

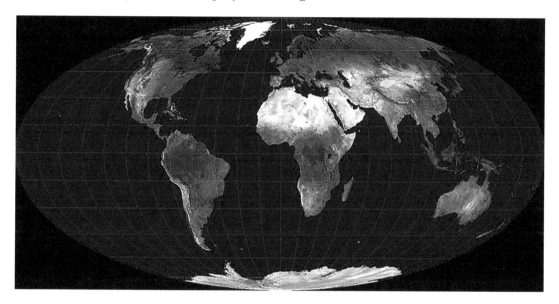

Figure 2-3. *The world seen in a Mollweide projection*
Source: `https://commons.wikimedia.org/wiki/File:Mollweide-`
`projection.jpg.` *Public Domain*

Example 2: Albers Equal Area Conic

The Albers equal area conic projection takes a very different approach, as it is conic. Making conic maps is often done to make some zones better represented. The Albers equal area conic projection, also called the Albert projection, projects the world to a two-dimensional map while respecting areas as shown in Figure 2-4.

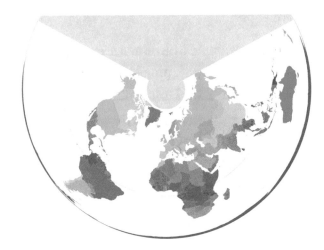

Figure 2-4. *The world seen in an Albers equal area conic projection*
Source: `https://commons.wikimedia.org/wiki/File:World_borders_`
`albers.png`*. Public Domain*

Conformal Projections

If shapes are important for your use case, you may want to use a conformal projection. Conformal projections are designed to preserve shapes. At the cost of distorting the areas on your map, this category of projections guarantees that all of the angles are preserved, and this makes sure that you see the "real" shapes on the map.

Mercator

The Mercator map is very well known, and it is the standard map projection for many projects. Its advantage is that it has north on top and south on the bottom while preserving local directions and shapes.

Unfortunately, locations far away from the equator are strongly inflated, for example, Greenland and Antarctica, while zones on the equator look too small in comparison (e.g., Africa).

The map looks like shown in Figure 2-5.

Figure 2-5. *The world seen in a Mercator projection*
Source: `https://commons.wikimedia.org/wiki/File:Mercator_projection_`
`of_world_with_grid.png.` *Public Domain*

Lambert Conformal Conic

The Lambert conformal conic projection is another conformal projection, meaning that it also respects local shapes. This projection is less widespread because it is a conic map, and conic maps have never become as popular as rectangles. However, it does just as well on plotting the earth while preserving shapes, and it has fewer problems with size distortion. It looks as shown in Figure 2-6.

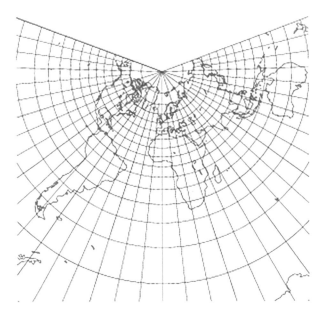

Figure 2-6. *The world seen in a Lambert conformal conic projection*
*Source: https://commons.wikimedia.org/wiki/File:Lambert_conformal_*
conical_projection_of_world_with_grid.png. Public Domain

Equidistant Projections

As the name indicates, you should use equidistant projections if you want a map that respects distances. In the two previously discussed projection types, there is no guarantee that distance between two points is respected. As you can imagine, this will be a problem for many use cases. Equidistant projections are there to save you if distances are key to your solution.

Azimuthal Equidistant Projection

One example of an equidistant projection is the azimuthal equidistant projection, also called Postel or zenithal equidistant. It preserves distances from the center and looks as shown in Figure 2-7.

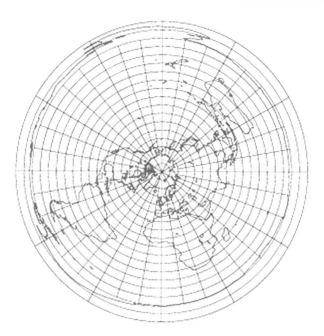

Figure 2-7. *The world seen in an azimuthal equidistant projection*
Source: `https://commons.wikimedia.org/wiki/File:Azimuthal_equidistant_`
`projection_of_world_with_grid.png.` *Public Domain*

Equidistant Conic Projection

The equidistant conic projection is another conic projection, but this time it preserves distance. It is also known as the simple conic projection, and it looks as shown in Figure 2-8.

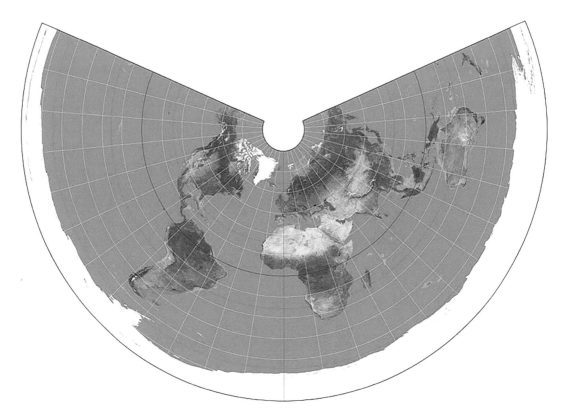

Figure 2-8. *The world seen in an equidistant conic projection*
Source: https://commons.wikimedia.org/wiki/File:Equidistant_conical_projection_of_world_with_grid.png. Public Domain

True Direction or Azimuthal Projections

Finally, azimuthal projections are designed to respect directions from one point to another. If, for example, you are making maps for navigation, azimuthal maps that respect directions may be essential. The direction of any point on the map will be guaranteed to be depicted in the right direction from the center.

An interesting additional advantage of this category is that they can be combined with the previous three types of maps, so they are not an either/or choice.

Lambert Equal Area Azimuthal

One example of an azimuthal projection is the Lambert equal area azimuthal. As the name indicates, it is not just azimuthal but also equal area. The world according to this projection looks as shown in Figure 2-9.

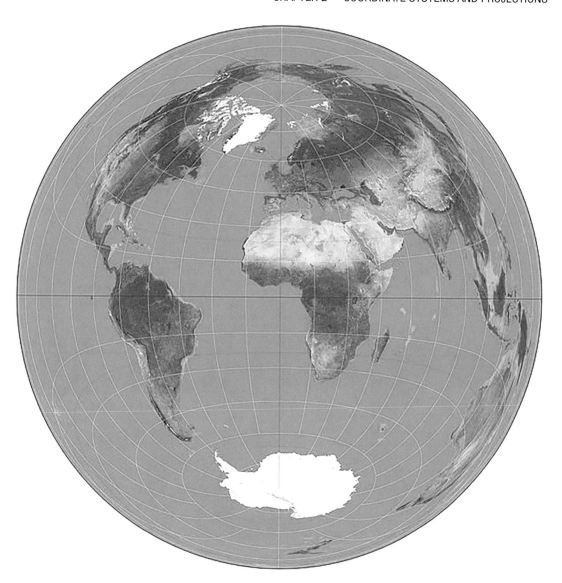

Figure 2-9. *The world seen in a Lambert equal area azimuthal projection*
Source: `https://commons.wikimedia.org/wiki/File:Lambert_azimuthal_` `equal-area_projection_of_world_with_grid.png.` *Public Domain*

Two-Point Equidistant Projection

Another azimuthal projection is the two-point equidistant projection, also called doubly equidistant projection. When making maps using this projection, one must choose two points on the map, and distances from those two points to any other point on the map are guaranteed to be at the same distance as the scale of the map. As an example, you can see in Figure 2-10 the two-point equidistant projection of Eurasia with two points from which all distances are respected. It is also azimuthal.

Figure 2-10. *The world seen in a two-point equidistant projection*
Source: `https://commons.wikimedia.org/wiki/File:World_borders_`
`donald.png.` *Public Domain*

Local Coordinate Systems

As you have seen, there are a number of Geographic Coordinate Systems, followed by an even larger number of map projections that each have their own specific mathematical definitions and are each created for a specific use case or practical and theoretical goals. What they have in common, however, is that they are generally made to function on the whole world.

A third type of coordinate systems is the Local Coordinate System. Local Coordinate Systems are, as you would expect, only suited for local maps and use cases. They are designed to function well on a smaller geographical zone. They often have an origin that is at a location that is logical for the zone but would not be logical for any other map. Therefore, they have coordinates only for the local zone and may be chosen such that the local, close-up maps depict the zone with as little distortion as possible.

We won't go further into this for now, as the limited use of most of those Local Coordinate Systems makes it hard to pick out interesting cases. Yet it is important to know that they exist, as you may encounter them, and you may even want to look up some Local Coordinate Systems for your personal area.

Which Coordinate System to Choose

After seeing this large variety and detail of coordinate systems, you may wonder which coordinate systems you should use. And, unfortunately, there is not one clear answer to this. Only if you have very clear needs, whether it is equal distances, equal shapes, or equal areas, there will be a very clear winner. Conic vs. cylindrical maps may also be a clear element of choice in this.

Another part of the answer is that you will very often be "forced" into working with a specific coordinate system, as you retrieve geodata datasets that are already in a chosen coordinate system. You can always go from one to the other, but if your dataset is large, conversions may take some time and pose problems: staying with a dataset's coordinate system can be the right choice if you have no specific needs.

Besides that, it can be a best practice to use "standard" choices. If all maps in your domain of application or in your region use a specific coordinate system, you may as well go with that choice for increased coherence.

One key takeaway here is that metadata is a crucial part of geodata. Sending datasets with coordinates while failing to mention details on the coordinate system used is very problematic. At the same time, if you are on the receiving end, stay critical of the data you receive, and pay close attention to whether or not you are on the right coordinate system. Mistakes are easily made and can be very impactful.

Playing Around with Some Maps

To end this chapter on coordinate systems, let's get to some practical applications. You will see how to create some data, change coordinate systems, and show some simple maps in those different coordinate systems. You will see a first example in which you create your own dataset using Google My Maps. You will project your data into another coordinate system and compare the differences.

Example: Working with Own Data

In the first example, you will learn an easy and accessible method for annotating geodata using Google My Maps. You will then import the map in Python and see how to convert it to a different coordinate system.

Step 1: Make Your Own Dataset on Google My Maps

Google My Maps is a great starting point for building simple geodatasets. Although it is not much used in professional use cases, it is a great way to get started or to make some quick sketches for smaller projects.

To start building a Google My Maps, you need to go to your Google Drive (drive. google.com). Inside your Google Drive, you click New, and you select New Google My Map. Once you do this, it will automatically open a Google Maps–like page in which you have some additional tools. You will see the toolbars top left.

Step 2: Add Some Features on Your Map

You can add features on your map by clicking the ⬇ icon for points and the ⬠ icon for lines and polygons. Let's try to make a polygon that contains the whole country of France, by clicking around its borders. You should end up with a gray polygon. For copyright reasons, it is not possible to reprint the map in Google format, but you can freely access it over here:

www.google.com/maps/d/edit?mid=1phChS9aNUukXKk2MwOQyXvksRk-
HTOdZ&usp=sharing

Step 3: Export Your Map As a .KML

By doing this, you have just created a very simple geodata dataset that contains one polygon. Now, to get this geodata into a different environment, let's export the data.

To do so, you go to the three little dots to the right of the name of your map, and you click Export Map, or Download KML, depending on your version and settings. You can then select to extract **only the layer rather than the entire map**. This way, you will end up with a dataset that has your own polygon in it. Also, select the option to **get a .KML rather than a .KMZ**.

Step 4: Import the .KML in Python

Now, let's see how we can get this map into Python. You can use the geopandas library together with the Fiona library to easily import a .KML map. If you don't have these libraries installed, you can use Code Block 2-1 if you're in a Jupyter notebook.

Code Block 2-1. Installing the libraries

```
!pip install fiona
!pip install geopandas
```

Then you can use the code in Code Block 2-2 to import your map and show the data that is contained within it.

Code Block 2-2. Importing the data

```
import fiona
import geopandas as gpd

gpd.io.file.fiona.drvsupport.supported_drivers['KML'] = 'rw'
kmlfile = gpd.read_file("the/path/to/the/exported/file.kml")
print(kmlfile)
```

You'll find that there is just one line in this dataframe and that it contains a polygon called France. Figure 2-11 shows this.

```
    Name Description                                              geometry
0   France                  POLYGON Z ((2.44446 51.28826 0.00000, -5.37780...
```

Figure 2-11. *The contents of the dataframe. Image by author*

We can inspect that polygon in more detail by extracting it from the dataframe using the code in Code Block 2-3.

Code Block 2-3. Extracting the geometry from the dataframe

```
print(kmlfile.loc[0,'geometry'])
```

You will see that the data of this polygon is a sequence of coordinates indicating the contours. This looks like Code Block 2-4.

Code Block 2-4. The resulting geometry, output of Code Block 2-3

```
POLYGON Z ((2.4444624 51.288264 0, -5.3778032 48.3696347 0, -1.0711626
46.315323 0, -2.0599321 43.3398321 0, 2.5103804 42.2434336 0, 7.7178999
43.7697753 0, 8.135380400000001 48.8924149 0, 2.4444624 51.288264 0))
```

Step 5: Plot the Map

Now that you understand how to define polygon-shaped map data as a text file, let's use simple Python functionality to make this into a map. This can be done very easily by using the plot method, as shown in Code Block 2-5.

Code Block 2-5. Plotting the map

```
import matplotlib.pyplot as plt
kmlfile.plot()
plt.title('Map in WGS 84')
```

The result of this code is shown in Figure 2-12.

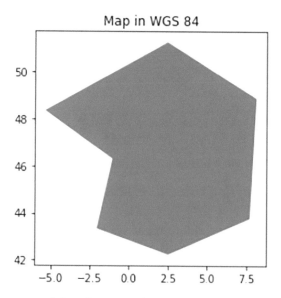

Figure 2-12. *The map resulting from Code Block 2-5. Image by author*

You will obtain a map of the polygon. You should recognize the exact shape of the polygon, as it was defined in your map or in the example map, depending on which one you used.

Step 6: Change the Coordinate System

KML standardly uses the WGS 84 Geographic Coordinate System. You can check that this is the case using the code in Code Block 2-6.

Code Block 2-6. Extracting the coordinate system

```
kmlfile.crs
```

You'll see the result like in Figure 2-13 being shown in your notebook.

```
<Geographic 2D CRS: EPSG:4326>
Name: WGS 84
Axis Info [ellipsoidal]:
- Lat[north]: Geodetic latitude (degree)
- Lon[east]: Geodetic longitude (degree)
Area of Use:
- name: World.
- bounds: (-180.0, -90.0, 180.0, 90.0)
Datum: World Geodetic System 1984 ensemble
- Ellipsoid: WGS 84
- Prime Meridian: Greenwich
```

Figure 2-13. *The result of Code Block 2-6. Image by author*

It may be interesting to see what happens when we plot the map into a very different coordinate system. Let's try to convert this map into a different coordinate system using the geopandas library. Let's change from the geographic WGS 84 into the projected Europe Lambert conformal conic map projection, which is also known as ESRI:102014.

The code in Code Block 2-7 makes the transformation from the source coordinate system to the target coordinate system.

Code Block 2-7. Changing the coordinate system

```
proj_kml = kmlfile.to_crs('ESRI:102014')
proj_kml
```

The result is shown in Figure 2-14.

Name	Description	geometry
0 France		POLYGON Z ((-518997.001 2420426.303 0.000, -11...

Figure 2-14. *The resulting dataframe from Code Block 2-7. Image by author*

Step 7: Plot the Map Again

Now to plot the polygon, the code in Code Block 2-8 will do the job.

Code Block 2-8. Plotting the map

```
proj_kml.plot()
plt.title('ESRI:102014 map')
```

46

The result is shown in Figure 2-15.

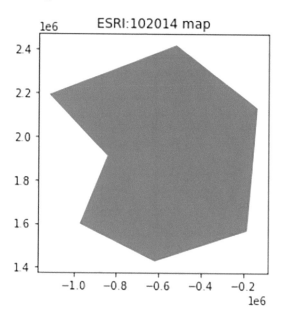

Figure 2-15.   *The plot resulting from Code Block 2-8. Image by author*

The coordinate systems have very different x and y values. To see differences in shape and size, you will have to look very closely. You can observe a slight difference in the way the angle on the left is made. The pointy bit on the left is pointing more toward the bottom in the left map, whereas it is pointing a bit more to the top in the right map. This is shown in Figure 2-16.

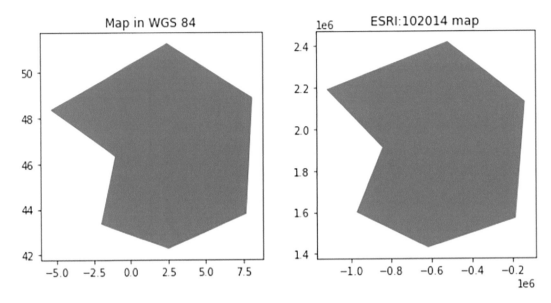

Figure 2-16. *Showing the two coordinate systems side by side. Image by author*

Although differences here are small, they can have a serious effect on your application. It is important to understand here that none of the maps are "wrong." They just use a different mathematical formula for projecting a 3D curved piece of land onto a 2D image.

Key Takeaways

1. Coordinate systems are mathematical descriptions of the earth that allow us to communicate about locations precisely

2. Many coordinate systems exist, and each has its own advantages and imperfections. One must choose a coordinate system depending on their use case.

3. Geographic Coordinate Systems use degrees and try to model the Earth as an ellipsoid or sphere.

4. Projected Coordinate Systems propose methods to convert the 3D reality onto a 2D map. This goes as the cost of some features of reality, which cannot be presented perfectly in 2D.

5.   There are a number of well-known projection categories.
     Equidistant makes sure that distances are not disturbed. Equal
     area projections make sure that areas are respected. Conformal
     projections maintain shape. Azimuthal projections keep
     directions the same.

6.   Local Coordinate Systems are interesting for very specific local
     studies or use cases, as they generally specify an origin and
     projection that makes sense on a local scale.

7.   You have seen how to annotate a geospatial dataset by hand
     using the easy-to-use and intuitive Google My Maps. You can
     export the map into a .KML format which can be easily imported
     into Python.

8.   You have seen how to import maps in Python using geopandas,
     and you have seen how to convert a geospatial dataset from one
     coordinate system to another. You have learned that the amount
     of difference will depend on the amount of difference between the
     different coordinate systems that you are using.

CHAPTER 3

Geodata Data Types

Throughout the previous chapters, you have been (secretly) exposed to a number of different geodata data types. In Chapter 1, we have talked about identifying points inside a coordinate system. In the previous chapter, you saw how a polygon of the shape of the country France was created. You have also seen examples of a TIFF data file being imported into Python.

Understanding geodata data types is key in working efficiently with geodata. In regular, tabular, datasets, it is generally not too costly to transform from one data type to another. Also, it is generally quite easy to say which data type is the "best" data type for a given variable or a given data point.

In geodata, the choice of data types is much more impacting. Transforming polygons of the shapes of countries into points is not a trivial task, as this would require defining (artificially) where you'd want to put each point. This would be in the "middle," which would often require quite costly computations.

The other way around, however, would not be possible anymore. Once you have a point dataset with the centers of countries, you would never be able to find the countries' exact boundaries anymore.

This problem is illustrated using the two images in Figures 3-1 and 3-2. You see one that has a map with the contours of the countries of the world, which has some black crosses indicating some of the countries' center points. In the second example, you see the same map, but with the polygons deleted. You can see clearly that once the polygon information is lost, you cannot go back to this information. This may be acceptable for some use cases, but it may be a problem for many use cases.

In this chapter, you will see the four main types of geodata data types, so that you will be comfortable working with all types, and you will be able to decide on the type of data to use.

© Joos Korstanje 2022
J. Korstanje, *Machine Learning on Geographical Data Using Python*,
https://doi.org/10.1007/978-1-4842-8287-8_3

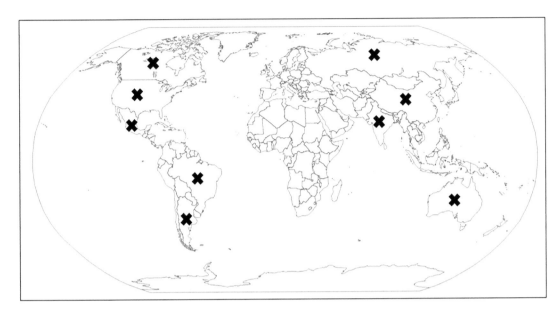

Figure 3-1. *Putting center points in polygons is possible*
Image adapted from geopandas (BSD 3 Clause License)

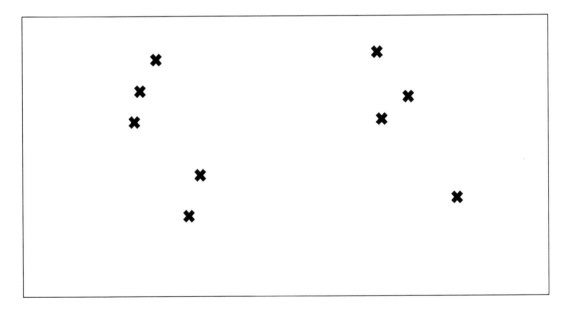

Figure 3-2. *Going from points back to polygons is not possible*
Image by author

Vector vs. Raster Data

The big split in geodata data types is between vector data and raster data. There is a fundamental difference in how those two are organized.

Vector data is data that contains objects with the specific coordinates of those objects. Those objects are points, lines, and polygons. In the two images introduced earlier, you have seen vector data. The first dataset is a polygon dataset that contains the spatial location data of those polygons in a given coordinate system. The second map shows a point dataset: it contains points and their coordinates.

A raster dataset works differently. Raster data is like a digital image. Just like images, a raster dataset is split into a huge number of very small squares: these are pixels. For every pixel, a value is stored.

You see that the big difference between vector and raster data is that vector data stores only objects with their coordinates. The rest of the map is effectively empty and that is not a problem in vector data. In the real world, this is how many objects work. If you think about a transportation dataset containing motorways and railroads, you can understand that most of the earth is **not** covered in them. It is much more efficient to just state where the objects are located (vector approach) than to state for each pixel on the world whether or not there is a road present (raster approach).

Raster data must have a value in every pixel, making it particularly useful for representing real-world phenomena that are not limited to a particular space. One good example is an elevation map: every location on earth has a specific height. By cutting the world into pixels, you could assign the height of each location (wrt sea level). This would give you a great map for relief mapping. Other continuous use cases exist like mapping temperature over the earth, mapping pollution values, and much more.

Dealing with Attributes in Vector and Raster

Coordinates are generally not the only thing that you want to know about your polygons or your raster. If you are collecting geodata, you are generally interested to know more about your locations than just where they are. Additional information is stored in what we call the attributes of your dataset.

As there is a fundamental difference in working with vector and raster data, it is interesting to understand how one would generally solve such data storage.

In the case of vector data, you will generally see a table that contains one row for each object (point, line, or polygon). In the columns of the table, you will see an ID and the geographic information containing the shape and coordinates. You can easily imagine adding to this any column of your choice with additional information about this object, like a name, or any other data that you may have about it: population size of the country, date of creation, and the list goes on. Figure 3-3 shows an example of this.

◢	A	B	C	D	E	F
1	object id	object type	coordinates	name	other_attribute_1	other_attribute_2
2	1	POLYGON	...	name1	...	...
3	2	POLYGON	...	name2	...	...
4	3	POLYGON	...	name3	...	...
5	4	POLYGON	...	name4	...	...
6	5	POLYGON	...	name5	...	...

Figure 3-3. *Vector data example. Image by author*

For raster data, the storage is generally image-like. As explained before, each pixel has a value. It is therefore common to store the data as a two-dimensional table in which each row represents a row of pixels, and each column represents a column of pixels. The values in your data table represent the values of one and only one variable. Working with raster data can be a bit harder to get into, as this image-like data format is not very accommodating to adding additional data. Figure 3-4 shows an example of this.

height	0	1	2	3	4
0	144	146	144	146	144
1	142	142	140	142	142
2	146	142	138	142	146
3	149	150	143	150	149
4	158	152	147	152	158

Figure 3-4. *Raster data. Image by author*

We will now get to an in-depth description of each of the data types that you are likely to encounter, and you will see how to work with them in Python.

Points

The simplest data type is probably the point. You have seen some examples of point data throughout the earlier chapters, and you have seen before that the point is one of the subtypes of vector data.

Points are part of vector data, as each point is an object on the map that has its own coordinates and that can have any number of attributes necessary. Point datasets are great for identifying locations of specific landmarks or other types of locations. Points cannot store anything like the shape of the size of landmarks, so it is important that you use points only if you do not need such information.

Definition of a Point

In mathematics, a point is generally said to be an exact location that has no length, width, or thickness. This is an interesting and important concept to understand about point data, as in geodata, the same is true.

A point consists only of one exact location, indicated by one coordinate pair (be it x and y, or latitude and longitude). Coordinates are numerical values, meaning that they can take an infinite number of decimals. The number 2.0, for example, is different than 2.1. Yet 2.01 is also different, 2.001 is a different location again, and 2.0001 is another, different, location.

Even if two points are very close to each other, it would theoretically not be correct that they are touching each other: as long as they are not in the same location, there will always be a small distance in between the points.

Another consideration is that if you have a point object, you cannot tell anything about its size. Although you could make points larger and smaller on the map, your point still stays at size 0. It is really just a location.

Importing an Example Point Dataset in Python

Let's start to work with a real-world point dataset in Python. For this data example, you'll use the 2018 Central Park Squirrel Census, an open dataset of the City of New York. It contains coordinates of squirrel sightings in the famous Central Park in New York. Each row of the dataset is a squirrel with its coordinates and a date and time of the sighting.

You can download the data from here: `https://data.cityofnewyork.us/ Environment/2018-Central-Park-Squirrel-Census-Squirrel-Data/vfnx-vebw`. If you click View data and then Export, you can download the file as a KML file, which you have already seen how to import in Python.

Before going into Python, it will be interesting to look at the text file through a text editor. If you open the KML file through a text editor, you'll see many blocks of data, each containing one squirrel sighting. You can use an XML formatter (just type XML formatter on Google for a free online one) and copy-paste a data block in it to make it more readable. You'll obtain something like Figure 3-5.

```
<ExtendedData>
    <Data name="x">
        <value>-73.9742811484852</value>
    </Data>
    <Data name="y">
        <value>40.775533619083</value>
    </Data>
    <Data name="unique_squirrel_id">
        <value>11B-PM-1014-08</value>
    </Data>
    <Data name="hectare">
        <value>11B</value>
    </Data>
    <Data name="shift">
        <value>PM</value>
    </Data>
    <Data name="date">
        <value>10142018</value>
    </Data>
    <Data name="hectare_squirrel_number">
        <value>8</value>
    </Data>
    <Data name="age">
        <value />
    </Data>
    <Data name="primary_fur_color">
        <value>Gray</value>
    </Data>
    <Data name="highlight_fur_color">
        <value />
    </Data>
```

Figure 3-5. *A screenshot of the content of the data. Image by author*
Data source: NYC OpenData. 2018 Central Park Squirrel Census

Of course, this is an extract, and the real list of variables about the squirrels is much longer. What is interesting to see is how the KML data format has stored point data just by having coordinates with it. Python (or any other geodata tool) will recognize the format and will be able to automatically import this the right way.

To import the data into Python, we can use the same code that was used in the previous chapter. It uses Fiona and geopandas to import the KML file into a geopandas dataframe. The code is shown in Code Block 3-1.

Code Block 3-1. Importing the Squirrel data

```
import fiona
import geopandas as gpd

gpd.io.file.fiona.drvsupport.supported_drivers['KML'] = 'rw'
kmlfile = gpd.read_file("2018 Central Park Squirrel Census - Squirrel
Data.kml")
print(kmlfile)
```

You will see the dataframe, containing geometry, being printed as shown in Figure 3-6.

```
      Name Description                        geometry
0                           POINT (-73.95613 40.79408)
1                           POINT (-73.96886 40.78378)
2                           POINT (-73.97428 40.77553)
3                           POINT (-73.95964 40.79031)
4                           POINT (-73.97027 40.77621)
...        ...           ...                        ...
3018                        POINT (-73.96394 40.79087)
3019                        POINT (-73.97040 40.78256)
3020                        POINT (-73.96659 40.78368)
3021                        POINT (-73.96399 40.78992)
3022                        POINT (-73.97548 40.76964)

[3023 rows x 3 columns]
```

Figure 3-6. *Capture of the Squirrel data. Image by author*
Data source: NYC OpenData. 2018 Central Park Squirrel Census

You can clearly see that each line is noted as follows: *POINT (coordinate coordinate)*. The coordinate system should be located in the geodataframe's attributes, and you can look at it using the code in Code Block 3-2.

Code Block 3-2. Inspecting the coordinate system

```
kmlfile.crs
```

You'll see the info about the coordinate system being printed, as shown in Figure 3-7.

```
<Geographic 2D CRS: EPSG:4326>
Name: WGS 84
Axis Info [ellipsoidal]:
- Lat[north]: Geodetic latitude (degree)
- Lon[east]: Geodetic longitude (degree)
Area of Use:
- name: World.
- bounds: (-180.0, -90.0, 180.0, 90.0)
Datum: World Geodetic System 1984 ensemble
- Ellipsoid: WGS 84
- Prime Meridian: Greenwich
```

Figure 3-7. *The output from Code Block 3-2. Image by author*
Data source: NYC OpenData. 2018 Central Park Squirrel Census

You can plot the map to see the squirrel sightings on the map using the code in Code Block 3-3. It is not very pretty for now, but additional visualization techniques will be discussed in Chapter 4. For now, let's focus on the data formats using Code Block 3-3.

Code Block 3-3. Plotting the data

```
import matplotlib.pyplot as plt
kmlfile.plot()
plt.title('Squirrels in Central Park (WGS84)')
```

You'll obtain the graph shown in Figure 3-8.

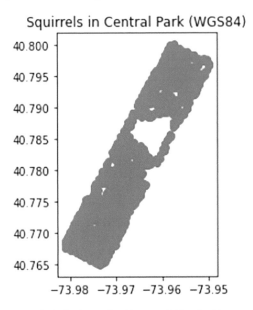

Figure 3-8. *Map of squirrel sightings in Central Park. Image by author Data source: NYC OpenData. 2018 Central Park Squirrel Census*

Some Basic Operations with Points

Let's try to execute some simple operations with this point dataset. This is a real-world dataset, so it will come with real-world issues. Let's get into it and see how to get to some results.

Filter Morning vs. Afternoon

As a study, let's try to do a filter to make one map for the morning observations and one for the afternoon observations. This may tell us whether there are any significant squirrel movements throughout the day.

If we look at the dataframe that was imported, you'll see that the Extended Variables part of our KML file was unfortunately not recognized by geopandas. Let's inspect the columns using the code in Code Block 3-4.

Code Block 3-4. Inspecting the columns

```
kmlfile.columns
```

You'll see that only the data shown in Figure 3-9 has been successfully imported.

```
Index(['Name', 'Description', 'geometry'], dtype='object')
```

Figure 3-9. *The output from Code Block 3-4. Image by author*

Now, this would be a great setback with any noncode geodata program, but as we are using Python, we have the full autonomy of finding a way to repair this problem. I am not saying that it is great that we have to parse the XML ourselves, but at least we are not blocked at this point.

XML parsing can be done using the xml library. XML is a tree-based data format, and using the xml element tree, you can loop through the different levels of the tree and go down in distance. Code Block 3-5 shows how to do this.

Code Block 3-5. Parsing XML data

```python
import xml.etree.ElementTree as ET
tree = ET.parse("2018 Central Park Squirrel Census - Squirrel Data.kml")
root = tree.getroot()

# loop through the xml to parse it data point by data point
df = []
for datapoint_i in range(1,3023):
  elementdata = root[0][1][datapoint_i][1]
  df_row = []

  for x in elementdata:
    df_row.append(x[0].text)

  df.append(df_row)

# get the column names
column_names = [x.attrib['name'] for x in elementdata]

# make into a dataframe and print
import pandas as pd
data = pd.DataFrame(df, columns = column_names)
data
```

You will end up with something like Figure 3-10.

	x	y	unique_squirrel_id	hectare	shift	date	hectare_squirrel_number	age
0	-73.9688574691102	40.7837825208444	21B-AM-1019-04	21B	AM	10192018	4	None
1	-73.9742811484852	40.775533619083	11B-PM-1014-08	11B	PM	10142018	8	None
2	-73.9596413903948	40.7903128889029	32E-PM-1017-14	32E	PM	10172018	14	Adult
3	-73.9702676472613	40.7762126854894	13E-AM-1017-05	13E	AM	10172018	5	Adult
4	-73.9683613516225	40.7725908847499	11H-AM-1010-03	11H	AM	10102018	3	Adult
...	...	...	...	...	...	...	...	...
3017	-73.9639431360458	40.7908677445466	30B-AM-1007-04	30B	AM	10072018	4	Adult
3018	-73.9704015859639	40.7825600069973	19A-PM-1013-05	19A	PM	10132018	5	Adult
3019	-73.9665871993517	40.7836775064883	22D-PM-1012-07	22D	PM	10122018	7	Adult
3020	-73.9639941227864	40.7899152327912	29B-PM-1010-02	29B	PM	10102018	2	None
3021	-73.9754794191553	40.7696404489025	5E-PM-1012-01	05E	PM	10122018	1	Adult

3022 rows × 30 columns

Figure 3-10. *The result of Code Block 3-5. Image by author*
Data source: NYC OpenData. 2018 Central Park Squirrel Census

We can now (finally) apply our filter on the column shift, using the code in Code Block 3-6.

Code Block 3-6. Apply the filter

```
AM_data = data[data['shift'] == 'AM']
PM_data = data[data['shift'] == 'PM']
```

To make the plots, we have to go back to a geodataframe again. This can be done by combining the variables x and y into a point geometry as shown in Code Block 3-7.

Code Block 3-7. Create geometry format

```
AM_geodata = gpd.GeoDataFrame(AM_data, geometry=gpd.points_from_xy(AM_
data['x'], AM_data['y']))
PM_geodata = gpd.GeoDataFrame(PM_data, geometry=gpd.points_from_xy(PM_
data['x'], PM_data['y']))
```

We finish by building the two plots using Code Block 3-8.

Code Block 3-8. Building the two plots

```
AM_geodata.plot()
plt.title('AM squirrels')

PM_geodata.plot()
plt.title('PM squirrels')
```

The result is shown in Figure 3-11. You now have the maps necessary to investigate differences in AM and PM squirrels. Again, visual parameters can be improved here, but that will be covered in Chapter 4. For now, we focus on the data types and their possibilities.

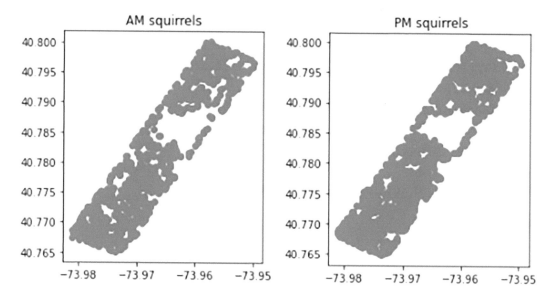

Figure 3-11. *The maps resulting from Code Block 3-8 Image by author Data source: NYC OpenData. 2018 Central Park Squirrel Census*

Lines

Line data is the second category of vector data in the world of geospatial data. They are the logical next step after points. Let's get into the definitions straight away.

Definition of a Line

Lines are also well-known mathematical objects. In mathematics, we generally consider straight lines that go from one point to a second point. Lines have no width, but they do have a length.

In geodata, line datasets contain not just one line, but many lines. Line segments are straight, and therefore they only need a from point and a to point. This means that a line needs two sets of coordinates (one of the first point and one of the second point).

Lines consist of multiple line segments, and they can therefore take different forms, consisting of straight line segments and multiple points. Lines in geodata can therefore represent the shape of features in addition to length.

An Example Line Dataset in Python

In this example, you will discover how to work with lines, by creating a line dataset from an open dataset that contains only coordinates. You can find two data files for this example at www.kaggle.com/usdot/flight-delays/data?select=flights.csv; it is a licensed public domain.

You can download two datasets: flights.csv and airports.csv. They are both CSV files with a .csv extension. You can import them easily into pandas using the code in Code Block 3-9 and Code Block 3-10.

Code Block 3-9. Import the flights data in Python

```
import pandas as pd
flights_data = pd.read_csv('flights.csv')
flights_data
```

The flights data is shown in Figure 3-12.

	YEAR	MONTH	DAY	DAY_OF_WEEK	AIRLINE	FLIGHT_NUMBER	TAIL_NUMBER	ORIGIN_AIRPORT	DESTINATION_AIRPORT	SCHEDULED_DEPARTURE	DEPARTURE_TIME	DEPARTURE_DELAY	TA
0	2015	1	1	4	AS	98	N407AS	ANC	SEA	5	2354 0	-11.0	
1	2015	1	1	4	AA	2336	N3KUAA	LAX	PBI	10	2 0	-8.0	
2	2015	1	1	4	US	840	N171US	SFO	CLT	20	18.0	-2.0	
3	2015	1	1	4	AA	258	N3HYAA	LAX	MIA	20	15 0	-5.0	
4	2015	1	1	4	AS	135	N527AS	SEA	ANC	25	24 0	-1.0	
...	...	...	...	...	...	...	...	...	...	...	...	...	
41536	2015	1	3	6	WN	1524	N787SA	PIT	MCO	1650	1707 0	17.0	
41537	2015	1	3	6	WN	100	N8621A	LAS	HOU	1650	1858 0	128.0	
41538	2015	1	3	6	WN	825	N654SW	LAS	MCI	1650	1706 0	16.0	
41539	2015	1	3	6	WN	527	N716SW	BWI	MSY	1650	1741 0	51.0	
41540	2015	1	3	6	WN	248	N642WN	DAL	SAN	1650	1802 0	72.0	

Figure 3-12. *The flights data. Image by author*
Data source: `www.kaggle.com/usdot/flight-delays`, *Public Domain*

Code Block 3-10. Importing the airports data in Python

```
geolookup = pd.read_csv('airports.csv')
geolookup
```

The airports data is shown in Figure 3-13.

	IATA_CODE	AIRPORT	CITY	STATE	COUNTRY	LATITUDE	LONGITUDE
0	ABE	Lehigh Valley International Airport	Allentown	PA	USA	40.65236	-75.44040
1	ABI	Abilene Regional Airport	Abilene	TX	USA	32.41132	-99.68190
2	ABQ	Albuquerque International Sunport	Albuquerque	NM	USA	35.04022	-106.60919
3	ABR	Aberdeen Regional Airport	Aberdeen	SD	USA	45.44906	-98.42183
4	ABY	Southwest Georgia Regional Airport	Albany	GA	USA	31.53552	-84.19447
...	...	...	...	...	...	...	...
317	WRG	Wrangell Airport	Wrangell	AK	USA	56.48433	-132.36982
318	WYS	Westerly State Airport	West Yellowstone	MT	USA	44.68840	-111.11764
319	XNA	Northwest Arkansas Regional Airport	Fayetteville/Springdale/Rogers	AR	USA	36.28187	-94.30681
320	YAK	Yakutat Airport	Yakutat	AK	USA	59.50336	-139.66023
321	YUM	Yuma International Airport	Yuma	AZ	USA	32.65658	-114.60597

Figure 3-13. *The airports data. Image by author*
Data source: `www.kaggle.com/usdot/flight-delays`, *Public Domain*

As you can see inside the data, the airports.csv is a file with geolocation information, as it contains the latitude and longitude of all the referenced airports. The flights. csv contains a large number of airplane routes in the USA, identified by origin and destination airport. Our goal is to convert the routes into georeferenced line data: a line with a from and to coordinate for each airplane route.

Let's start by converting the latitude and longitude variables into a point, so that the geometry can be recognized in further operations. The following code loops through the rows of the dataframe to generate a new variable. The whole operation is done twice, as to generate a "to/destination" lookup dataframe and a "from/source" lookup dataframe. This is shown in Code Block 3-11.

Code Block 3-11. Converting the data – part 1

```
# convert coordinates of geo lookup to a Point
# make to and from data set for to and from join
from shapely.geometry import Point
from_geo_lookup = geolookup[['IATA_CODE', 'LATITUDE', 'LONGITUDE']]
from_geo_lookup['geometry_from']= [Point(x,y) for x,y in zip( from_geo_
lookup['LONGITUDE'], from_geo_lookup['LATITUDE'])]
from_geo_lookup = from_geo_lookup[['IATA_CODE','geometry_from' ]]

to_geo_lookup = geolookup[['IATA_CODE', 'LATITUDE', 'LONGITUDE']]
to_geo_lookup['geometry_to']= [Point(x,y) for x,y in zip( to_geo_
lookup['LONGITUDE'], to_geo_lookup['LATITUDE'])]
to_geo_lookup = to_geo_lookup[['IATA_CODE','geometry_to' ]]
```

As the data types are not aligned, the easiest hack here is to convert all the numbers to strings. There are some missing codes and this would be better to solve by inspecting the data quality issues, but for this introductory example, the string conversion does the job for us. You can also see that some columns are dropped here. This is done in Code Block 3-12.

Code Block 3-12. Converting the data – part 2

```
# align data types for the joins
from_geo_lookup['IATA_CODE'] = from_geo_lookup['IATA_CODE'].map(str)
to_geo_lookup['IATA_CODE'] = to_geo_lookup['IATA_CODE'].map(str)

flights_data['ORIGIN_AIRPORT'] = flights_data['ORIGIN_AIRPORT'].map(str)
flights_data['DESTINATION_AIRPORT'] = flights_data['DESTINATION_AIRPORT'].
map(str)
flights_data = flights_data[['ORIGIN_AIRPORT', 'DESTINATION_AIRPORT']]
```

We now get to the step to merge the dataframes of the flights together with the from and to geographical lookups that we just created. The code in Code Block 3-13 merges two times (once with the from coordinates and once with the to coordinates).

Code Block 3-13. Merging the data

```
flights_data = flights_data.merge(from_geo_lookup, left_on = 'ORIGIN_
AIRPORT', right_on = 'IATA_CODE')
flights_data = flights_data.merge(to_geo_lookup, left_on = 'DESTINATION_
AIRPORT', right_on = 'IATA_CODE')
flights_data = flights_data[['geometry_from', 'geometry_to']]
flights_data
```

After running this code, you will end up with a dataframe that still contains one row per route, but it has now got two georeference columns: the from coordinate and the to coordinate. This result is shown in Figure 3-14.

	geometry_from	geometry_to
0	POINT (-149.99619 61.17432)	POINT (-122.30931 47.44898)
1	POINT (-149.99619 61.17432)	POINT (-122.30931 47.44898)
2	POINT (-149.99619 61.17432)	POINT (-122.30931 47.44898)
3	POINT (-149.99619 61.17432)	POINT (-122.30931 47.44898)
4	POINT (-149.99619 61.17432)	POINT (-122.30931 47.44898)
...	...	...
41536	POINT (-131.71374 55.35557)	POINT (-132.36982 56.48433)
41537	POINT (-131.71374 55.35557)	POINT (-132.36982 56.48433)
41538	POINT (-132.94528 56.80165)	POINT (-132.36982 56.48433)
41539	POINT (-132.94528 56.80165)	POINT (-132.36982 56.48433)
41540	POINT (-132.94528 56.80165)	POINT (-132.36982 56.48433)

41541 rows × 2 columns

Figure 3-14. *The dataframe resulting from Code Block 3-13. Image by author Data source:* www.kaggle.com/usdot/flight-delays, *Public Domain*

The final step of the conversion process is to make lines out of this to and from points. This can be done using the LineString function as shown in Code Block 3-14.

Code Block 3-14. Convert points to lines

```
# convert points to lines
from shapely.geometry import LineString

lines = []
for i,row in flights_data.iterrows():
  try:
    point_from = row['geometry_from']
    point_to = row['geometry_to']
    lines.append(LineString([point_from, point_to]))
```

```
except:
  #some data lines are faulty so we ignore them
  pass
```

```
geodf = gpd.GeoDataFrame(lines, columns=['geometry'])
geodf
```

You will end up with a new geometry variable that contains only LINESTRINGS. Inside each LINESTRING, you see the four values for the two coordinates (x and y from, and x and y to). This is shown in Figure 3-15.

	geometry
0	LINESTRING (-149.99619 61.17432, -122.30931 47..
1	LINESTRING (-149.99619 61.17432, -122.30931 47..
2	LINESTRING (-149.99619 61.17432, -122.30931 47..
3	LINESTRING (-149.99619 61.17432, -122.30931 47..
4	LINESTRING (-149.99619 61.17432, -122.30931 47..
...	..
41536	LINESTRING (-131.71374 55.35557, -132.36982 56..
41537	LINESTRING (-131.71374 55.35557, -132.36982 56..
41538	LINESTRING (-132.94528 56.80165, -132.36982 56..
41539	LINESTRING (-132.94528 56.80165, -132.36982 56..
41540	LINESTRING (-132.94528 56.80165, -132.36982 56..

Figure 3-15. *Linestring geometry. Image by author*
Data source: `www.kaggle.com/usdot/flight-delays`, *Public Domain*

Now that you have created your own line dataset, let's make a quick visualization as a final step. As before, you can simply use the plot functionality to generate a basic plot of your lines. This is shown in Code Block 3-15.

Code Block 3-15. Plot the data

```
# plot the lines
import matplotlib.pyplot as plt
geodf.plot(figsize=(12,12))
plt.title('the world as airpline trajectory lines')
```

You should now obtain the map of the USA given in Figure 3-16. You clearly see all the airplane trajectories expressed as straight lines. Clearly, not all of it is correct as flights do not take a straight line (as seen in a previous chapter). However, it gives a good overview of how to work with line data, and it is interesting to see that we can even recognize the USA map by just using flight lines (with some imagination).

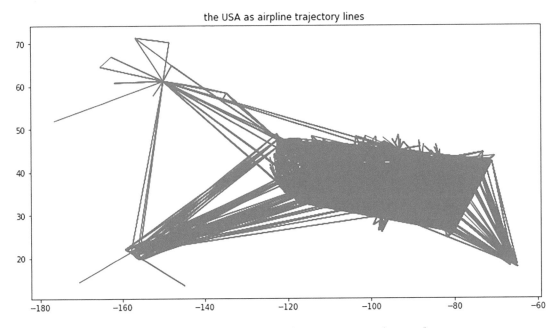

Figure 3-16. *Plot resulting from Code Block 3-15. Image by author* *Data source:* www.kaggle.com/usdot/flight-delays, *Public Domain*

Polygons

Polygons are the next step in complexity after points and lines. They are the third and last category of vector geodata.

Definition of a Polygon

In mathematics, polygons are defined as two-dimensional shapes, made up of lines that connect to make a closed shape. Examples are triangles, rectangles, pentagons, etc. A circle is not officially a polygon as it is not made up of straight lines, but you could imagine a lot of very small straight lines being able to approximate a circle relatively well.

In geodata, the definition of the polygon is not much different. It is simply a list of points that together make up a closed shape. Polygons are generally a much more realistic representation of the real world. Landmarks are often identified by points, but as you get to a very close-up map, you would need to represent the landmark as a polygon (the contour) to be useful. Roads could be well represented by lines (remember that lines have no width) but would have to be replaced by polygons once the map is at a small enough scale to see houses, roads, etc.

Polygons are the data type that has the most information as they are able to store location (just like points and lines), length (just like lines), and also area and perimeter.

An Example Polygon Dataset in Python

For this example, you can download a map of countries directly through geopandas. You can use the code in Code Block 3-16 to import the data in Python using geopandas.

Code Block 3-16. Reading polygon data

```
import geopandas as gpd
geojsonfile = gpd.read_file(gpd.datasets.get_path('naturalearth_lowres'))
print(geojsonfile)
```

You'll see the content of the polygon dataset in Figure 3-17. It contains some polygons and some multipolygons (polygons that consist of multiple polygons, e.g., the USA has Alaska that is not connected to their other land, so they need multiple polygons to describe their territory).

	pop_est	continent	name	iso_a3	gdp_md_est	geometry
0	920938	Oceania	Fiji	FJI	8374.0	MULTIPOLYGON (((180.00000 -16.06713, 180.00000...
1	53950935	Africa	Tanzania	TZA	150600.0	POLYGON ((33.90371 -0.95000, 34.07262 -1.05982...
2	603253	Africa	W. Sahara	ESH	906.5	POLYGON ((-8.66559 27.65643, -8.66512 27.58948...
3	35623680	North America	Canada	CAN	1674000.0	MULTIPOLYGON (((-122.84000 49.00000, -122.9742...
4	326625791	North America	United States of America	USA	18560000.0	MULTIPOLYGON (((-122.84000 49.00000, -120.0000...
...	...	...	...	...	...	...
172	7111024	Europe	Serbia	SRB	101800.0	POLYGON ((18.82982 45.90887, 18.82984 45.90888...
173	642550	Europe	Montenegro	MNE	10610.0	POLYGON ((20.07070 42.58863, 19.80161 42.50009...
174	1895250	Europe	Kosovo	-99	18490.0	POLYGON ((20.59025 41.85541, 20.52295 42.21787...
175	1218208	North America	Trinidad and Tobago	TTO	43570.0	POLYGON ((-61.68000 10.76000, -61.10500 10.890...
176	13026129	Africa	S. Sudan	SSD	20880.0	POLYGON ((30.83385 3.50917, 29.95350 4.17370, ...

177 rows × 6 columns

Figure 3-17. *Content of the polygon dataImage by author*
Source: geopandas, BSD 3 Clause Licence

You can easily create a map, as we did before, using the plot function. This is demonstrated in Code Block 3-17, and this time, it will automatically plot the polygons.

Code Block 3-17. Plotting polygons

```
geojsonfile.plot()
```

The plot looks as shown in Figure 3-18.

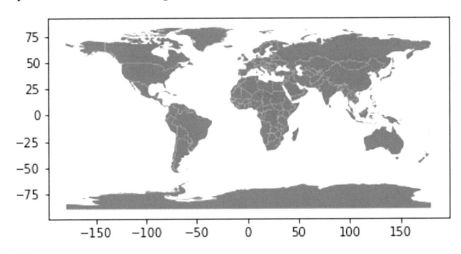

Figure 3-18. *The plot of polygons as created in Code Block 3-17. Image by author*
Source: geopandas, BSD 3 Clause Licence

Some Simple Operations with Polygons

As polygons are more complex shapes than the previously seen data types, let's check out some things that we can do with them. One thing that we can do with polygons is to compute their surface area very easily. For example, let's compute a list of the ten smallest countries.

To do this, we first compute the area of all the countries, using the area attribute of the polygons. We then sort by area and show the first ten lines. The code is shown in Code Block 3-18.

Code Block 3-18. Working with the area

```
geojsonfile['area'] = geojsonfile['geometry'].apply(lambda x: x.area)
geojsonfile.sort_values('area').head(10)
```

In Figure 3-19, you'll see the first ten rows of this data, which are the world's smallest countries in terms of surface area.

	pop_est	continent	name	iso_a3	gdp_md_est	geometry	area
128	594130	Europe	Luxembourg	LUX	58740.00	POLYGON ((6.04307 50.12805, 6.24275 49.90223, ...	0.301516
160	265100	Asia	N. Cyprus	-99	3600.00	POLYGON ((32.73178 35.14003, 32.80247 35.14550...	0.374644
79	4543126	Asia	Palestine	PSE	21220.77	POLYGON ((35.39756 31.48909, 34.92741 31.35344...	0.480314
161	1221549	Asia	Cyprus	CYP	29260.00	POLYGON ((32.73178 35.14003, 32.91957 35.08783...	0.613351
89	282814	Oceania	Vanuatu	VUT	723.00	MULTIPOLYGON (((167.21680 -15.89185, 167.84488...	0.631326
175	1218208	North America	Trinidad and Tobago	TTO	43570.00	POLYGON ((-61.68000 10.76000, -61.10500 10.890...	0.639000
45	3351827	North America	Puerto Rico	PRI	131000.00	POLYGON ((-66.28243 18.51476, -65.77130 18.426...	0.788009
149	443593	Asia	Brunei	BRN	33730.00	POLYGON ((115.45071 5.44773, 115.40570 4.95523...	0.872053
77	6229794	Asia	Lebanon	LBN	85160.00	POLYGON ((35.82110 33.27743, 35.55280 33.26427...	0.983810
85	2314307	Asia	Qatar	QAT	334500.00	POLYGON ((50.81011 24.75474, 50.74391 25.48242...	1.015639

Figure 3-19. *The first ten rows of the data. Image by author*
Source: geopandas, BSD 3 Clause Licence

We can also compute the length of the borders by calculating the length of the polygon borders. The length attribute allows us to do so. You can use the code in Code Block 3-19 to identify the ten countries with the longest contours.

Code Block 3-19. Identify the ten countries with longest contours

```
geojsonfile['length'] = geojsonfile['geometry'].apply(lambda x: x.length)
geojsonfile.sort_values('length', ascending=False).head(10)
```

You'll see the result in Figure 3-20, with Antarctica being the winner. Attention though, as this may be distorted by coordinate system choice. You may remember that some commonly used coordinate systems have strong distortions toward the poles and make more central locations smaller. This could influence the types of computations that are being done here. If a very precise result is needed, you'd need to tackle this question, but for a general idea of the countries with the longest borders, the current approach will do.

	pop_est	continent	name	iso_a3	gdp_md_est	geometry	area	length
159	4050	Antarctica	Antarctica	ATA	810.0	MULTIPOLYGON (((-48.66062 -78.04702, -48.15140...	6028.836194	1041.993521
3	35623680	North America	Canada	CAN	1674000.0	MULTIPOLYGON (((-122.84000 49.00000, -122.9742...	1712.995228	916.062855
18	142257519	Europe	Russia	RUS	3745000.0	MULTIPOLYGON (((178.72530 71.09880, 180.00000 ...	2935.205205	766.391129
4	326625791	North America	United States of America	USA	18560000.0	MULTIPOLYGON (((-122.84000 49.00000, -120.0000...	1122.281921	356.977119
22	57713	North America	Greenland	GRL	2173.0	POLYGON ((-46.76379 82.62796, -43.40644 83.225...	677.509565	235.678216
139	1379302771	Asia	China	CHN	21140000.0	MULTIPOLYGON (((109.47521 18.19770, 108.65521 ...	954.635341	221.976583
8	260580739	Asia	Indonesia	IDN	3028000.0	MULTIPOLYGON (((141.00021 -2.60015, 141.01706 ...	148.135821	215.431767
137	23232413	Oceania	Australia	AUS	1189000.0	MULTIPOLYGON (((147.68926 -40.80826, 148.28907...	695.545501	162.605664
29	207353391	South America	Brazil	BRA	3081000.0	POLYGON ((-53.37366 -33.76838, -53.65054 -33.2...	710.185243	158.445684
21	5320045	Europe	Norway	-99	364700.0	MULTIPOLYGON (((15.14282 79.67431, 15.52255 80...	90.496255	127.160983

Figure 3-20. *Dataset resulting from Code Block 3-19. Image by author*

Rasters/Grids

Raster data, also called grid data, is the counterpart of vector data. If you're used to working with digital images in Python, you might find raster data quite similar. If you're used to working with dataframes, it may be a bit more abstract, and take a moment to get used to it.

Definition of a Grid or Raster

A grid, or a raster, is a network of evenly spaced horizontal and vertical lines that cut a space into small squares. In images, we tend to call each of those squares a pixel. In mathematics, there are many other uses for grids, so pixel is not a universal term.

Grid geodata are grids that contain one value per "pixel," or cell, and therefore they end up being a large number of values filling up a square box. If you are not familiar with this approach, it may seem unlikely that this could be converted into something, but once a color scale is assigned to the values, this is actually how images are made. For raster/grid maps, the same is true.

Importing a Raster Dataset in Python

On the following website, you can download a GeoTIF file that contains an interpolated terrain model of Kerbernez in France:

```
https://geo.data.gouv.fr/en/datasets/b0a420b9e003d45aaf0670446f0d600df14430cb
```

You can use the code in Code Block 3-20 to read and show the raster file in Python.

Code Block 3-20. Opening the raster data

```
import rasterio
griddata = r'ore-kbz-mnt-litto3d-5m.tif'
img = rasterio.open(griddata)
matrix = img.read()
matrix
```

As you can see in Figure 3-21, this data looks nothing like a geodataframe whatsoever. Rather, it is just a matrix full of the values of the one (and only one) variable that is contained in this data.

```
array([[[56.  , 55.96, 56.06, ...,  0.38,  0.44,  0.56],
        [56.04, 56.05, 56.  , ...,  0.31,  0.43,  0.42],
        [56.  , 55.99, 56.02, ...,  0.38,  0.51,  0.37],
        ...,
        [55.68, 55.83, 55.88, ..., 37.67, 37.72, 37.78],
        [55.69, 55.74, 55.78, ..., 37.71, 37.76, 37.83],
        [55.9 , 55.77, 55.92, ..., 37.76, 37.79, 37.87]]])
```

Figure 3-21. *The raster data in Python. Image by author*
Data source: Ministry of DINSIC, https://geo.data.gouv.fr/en/datasets/
b0a420b9e003d45aaf0670446f0d600df14430cb. Creation data: Unknown.
Open Licence 2.0: www.etalab.gouv.fr/wp-content/uploads/2018/11/open-
licence.pdf

You can plot this data using the default color scale, and you will see what this numerical representation actually contains. As humans, we are particularly bad at reading and interpreting something from a large matrix like the one earlier, but when we see it color-coded into a map, we can get a much better feeling of what we are looking at. The code in Code Block 3-21 does exactly that.

Code Block 3-21. Plotting the raster data

```
from rasterio.plot import show
show(img)
```

The result of this is shown in Figure 3-22.

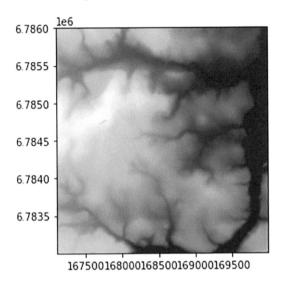

Figure 3-22. *The plot of the raster data. Image by author*
Data source: Ministry of DINSIC, https://geo.data.gouv.fr/en/datasets/
b0a420b9e003d45aaf0670446f0d600df14430cb. Creation data: Unknown.
Open Licence 2.0: www.etalab.gouv.fr/wp-content/uploads/2018/11/open-
licence.pdf

Raster data is a bit more limited than vector data in terms of adding data to it. Adding more variables would be quite complex, except for making the array into a 3D, where the third dimension contains additional data. However, for plotting, this would not be of any help, as the plot color would still be one color per pixel, and you could never show multiple variables for each pixel with this approach.

Raster data is still a very important data type that you will often need and often use. Any value that needs to be measured over a large area will be more suitable to raster. Examples like height maps, pollution maps, density maps, and much more are all only solvable with rasters. Raster use cases are generally a bit more mathematically complex, as they often use a lot of matrix computations. You'll see examples of these mathematical operations throughout the later chapters of the book.

Key Takeaways

1. There are two main categories of geodata: vector and raster. They have fundamentally different ways of storing data.

2. Vector data stores objects and stores the geospatial references for those objects.

3. Raster data cuts an area into equal-sized squares and stores a data value for each of those squares.

4. There are three main types of vector data: point, line, and polygon.

5. Points are zero-dimensional, and they have no size. They are only indicated by a single x,y coordinate. Points are great for indicating the location of objects.

6. Lines are one-dimensional. They have a length, but no width. They are indicated by two or more points in a sequence. Lines are great for indicating line-shaped things like rivers and roads.

7. Polygons are two-dimensional objects. They have a shape and size. Polygons are great when your objects are polygons and when you need to retain this information. Polygons can indicate the location of objects if you also need to locate their contour. It can also apply for rivers and roads when you also need to store data about their exact shape and width. Polygons are the data type that can retain the largest amount of information among the three vector data types.

8. Raster data is suitable for measurements that are continuous over an area, like height maps, density maps, heat maps, etc.

CHAPTER 4

Creating Maps

Mapmaking is one of the earliest and most obvious use cases of the field of geodata. Maps are a special form of data visualization: they have a lot of standards and are therefore easily recognizable and interpretable for almost anyone.

Just like other data visualization methods, maps are a powerful tool to share a message about a dataset. Visualization tools are often wrongly interpreted as an objective depiction of the truth, whereas in reality, map makers and visualization builders have a huge power of putting things on the map or leaving things out.

An example is color scale picking on maps. People are so familiar with some visualization techniques that when they see them, they automatically believe them.

Imagine a map showing pollution levels in a specific region. If you would want people to believe that pollution is not a big problem in the area, you could build and share a map that shows areas with low pollution as dark green and very strongly polluted areas as light green. Add to that a small, unreadable, legend, and people will easily interpret that there is no big pollution problem.

If you want to argue the other side, you could publish an alternative map that shows the exact same values, but you depict strong pollution as dark red and slight pollution as light red. When people see this map, they will directly be tempted to conclude that pollution is a huge problem in your area and that it needs immediate action.

It is important to understand that there is no truth in choosing visualization. There are however a number of levers in mapmaking that you should master well in order to create maps for your specific purpose. Whether your purpose is making objective maps, beautiful maps, or communicating a message, there are a number of tools and best practices that you will discover in this chapter. Those are important to remember when making maps and will come in handy when interpreting maps as well.

© Joos Korstanje 2022
J. Korstanje, *Machine Learning on Geographical Data Using Python*,
https://doi.org/10.1007/978-1-4842-8287-8_4

Mapping Using Geopandas and Matplotlib

As you have already seen numerous examples with geopandas throughout the earlier chapters of this book, the easiest way to start mapping is to work with geopandas in combination with matplotlib. Matplotlib is the standard plotting and visualization library in Python, and it has great and intuitive integration with geopandas.

Getting a Dataset into Python

For this first example, we'll be using an example that is based on the geopandas documentation (`https://geopandas.org/en/stable/docs/user_guide/mapping.html`). They have a rich documentation with a lot of detail on how to make maps using geopandas with matplotlib, so I recommend checking that out for more detailed mapping options that you may be interested in for your own use cases and visualizations.

To get started with this example, you can use the built-in example map from geopandas. It contains polygons of the world's countries. You can import it into Python with the code in Code Block 4-1.

Code Block 4-1. Importing the data

```
import geopandas as gpd
world = gpd.read_file(gpd.datasets.get_path("naturalearth_lowres"))
world.head()
```

Once you execute this code, you'll see the first five lines of the geodataframe containing the world's countries, as displayed in Figure 4-1.

	pop_est	continent	name	iso_a3	gdp_md_est	geometry
0	920938	Oceania	Fiji	FJI	8374.0	MULTIPOLYGON (((180.00000 -16.06713, 180.00000...
1	53950935	Africa	Tanzania	TZA	150600.0	POLYGON ((33.90371 -0.95000, 34.07262 -1.05982...
2	603253	Africa	W. Sahara	ESH	906.5	POLYGON ((-8.66559 27.65643, -8.66512 27.58948...
3	35623680	North America	Canada	CAN	1674000.0	MULTIPOLYGON (((-122.84000 49.00000, -122.9742...
4	326625791	North America	United States of America	USA	18560000.0	MULTIPOLYGON (((-122.84000 49.00000, -120.0000...

Figure 4-1. *The data. Image by author*
Data source: geopandas, BSD 3 Clause Licence

For this example, we'll make a map that is color-coded: colors will be based on the area of the countries. To get there, we need to add a column to the geodataframe that contains the countries' areas. This can be obtained using Code Block 4-2.

Code Block 4-2. Computing the areas

```
world['area'] = world.geometry.apply(lambda x: x.area)
world.head()
```

If you now look at the dataframe again, you'll see that an additional column is indeed present, as shown in Figure 4-2. It contains the area of each country and will help us in the mapmaking process.

	pop_est	continent	name	iso_a3	gdp_md_est	geometry	gdp_per_cap	area
0	920938	Oceania	Fiji	FJI	8374.0	MULTIPOLYGON (((180.00000 -16.06713, 180.00000...	0.009093	1.639511
1	53950935	Africa	Tanzania	TZA	150600.0	POLYGON ((33.90371 -0.95000, 34.07262 -1.05982...	0.002791	76.301964
2	603253	Africa	W. Sahara	ESH	906.5	POLYGON ((-8.66559 27.65643, -8.66512 27.58948...	0.001503	8.603984
3	35623680	North America	Canada	CAN	1674000.0	MULTIPOLYGON (((-122.84000 49.00000, -122.9742...	0.046991	1712.995228
4	326625791	North America	United States of America	USA	18560000.0	MULTIPOLYGON (((-122.84000 49.00000, -120.0000...	0.056823	1122.281921

Figure 4-2. *The data with an additional column. Image by author*
Data source: geopandas, BSD 3 Clause Licence

Making a Basic Plot

In previous chapters, you have already seen how to use the plot method on a geodataframe. You can add the pyplot rcParams to have a larger output size in your notebook, which I recommend as the standard size is fairly small. This is done in Code Block 4-3.

Code Block 4-3. Adding a figsize

```
import matplotlib.pyplot as plt
plt.rcParams["figure.figsize"] = [16,9]
world.plot()
```

If you do this, you'll obtain a plot that just contains the polygons, just like in Figure 4-3. There is no additional color-coding going on.

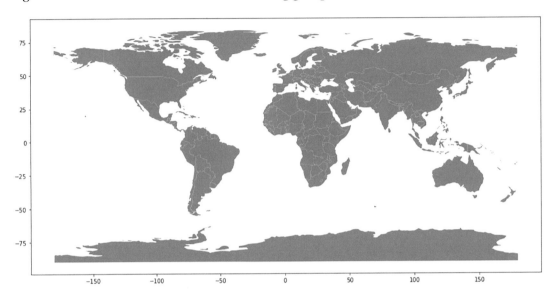

Figure 4-3. *Larger plot size. Image by author*
Data source: geopandas, BSD 3 Clause Licence

As the goal of our exercise is to color-code countries based on their total area, we'll need to start improving on this map with additional plotting parameters.

Adding color-coding to a plot is fairly simple using geopandas and matplotlib. The plot method can take an argument column, and when specifying a column name there, the map will automatically be color-coded based on this column.

In our example, we want to color-code with the newly generated variable called area, so we'll need to specify column='area' in the plot arguments. This is done in Code Block 4-4.

Code Block 4-4. Adding a color-coded column

```
world.plot(column='area', cmap='Greys')
```

You will see the black and white coded map as shown in Figure 4-4.

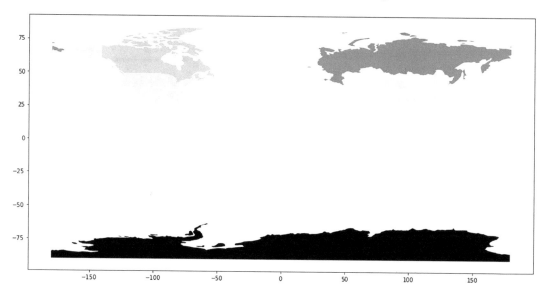

Figure 4-4. *The grayscale map resulting from Code Block 4-4. Image by author Data source: geopandas, BSD 3 Clause Licence*

Plot Title

Let's continue working on this map a bit more. One important thing to add to any visualization, including maps, is a title. A title will allow readers to easily understand what the goal of your map is.

When making maps with geopandas and matplotlib, you can use the matplotlib command plt.title to easily add a title on top of your map. The example in Code Block 4-5 shows you how it's done.

Code Block 4-5. Adding a plot title

```
world.plot(column='area', cmap='Greys')
plt.title('Area per country')
```

You will obtain the map in Figure 4-5. It is still the same map as before, but now has a title on top of it.

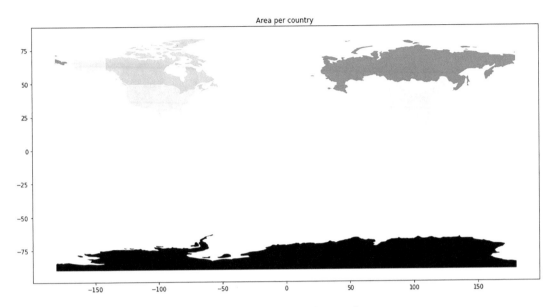

Figure 4-5. *The same map with a title. Image by author*
Data source: geopandas, BSD 3 Clause Licence

Plot Legend

Another essential part of maps (and other visualizations) is to add a legend whenever you use color or shape encodings. In our map, we are using color-coding to show the area of the countries in a quick visual manner, but we have not yet added a legend. It can therefore be confusing for readers of the map to understand which values are high areas and which indicate low areas.

In the code in Code Block 4-6, the plot method takes two additional arguments. Legend is set to True to generate a legend. The legend_kwds takes a dictionary with some additional parameters for the legend. The label will be the label of the legend, and the orientation is set to horizontal to make the legend appear on the bottom rather than on the side. A title is added at the end of the code, just like you saw in the previous part.

Code Block 4-6. Adding a legend

```
import matplotlib.pyplot as plt
world.plot(column='area', cmap='Greys', legend=True)
plt.title('Area per country')
```

You will obtain the plot in Figure 4-6.

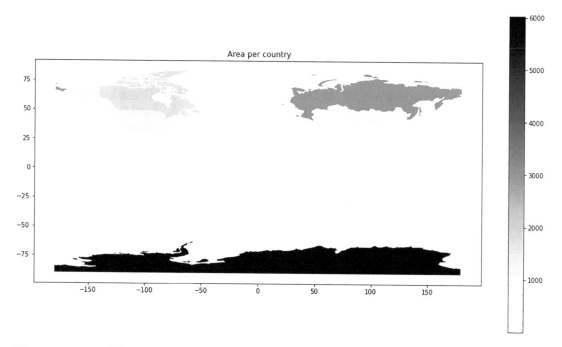

Figure 4-6. *Adding a legend to the map. Image by author*
Data source: geopandas, BSD 3 Clause Licence

This is the final version of this map for the current example. The map does a fairly good job at representing a numerical value for different countries. This type of use case is easily solvable with geopandas and matplotlib. Although it may not be the most aesthetically pleasing map, it is perfect for analytical purposes and the like.

Mapping a Point Dataset with Geopandas and Matplotlib

In the previous example, you have seen how to make a plot with a polygon dataset. In the current example, we'll go deeper into plotting with geopandas and matplotlib. You'll see how to take a second dataset from the built-in geopandas dataset, this time a point dataset, and plot it on top of a polygon dataset. This will teach you how to plot point datasets and at the same time how to plot multiple datasets on top of each other.

You can import the built-in dataset with the code shown in Code Block 4-7. It contains the cities of earth and has a point geometry. This is done in Code Block 4-7.

Code Block 4-7. Importing the data

```
cities = gpd.read_file(gpd.datasets.get_path('naturalearth_cities'))
cities.head()
```

When executing this code, you'll see the first five lines of the dataframe, just like shown in Figure 4-7. The column geometry shows the points, which are two coordinates just like you have seen in earlier chapters.

	name	geometry
0	Vatican City	POINT (12.45339 41.90328)
1	San Marino	POINT (12.44177 43.93610)
2	Vaduz	POINT (9.51667 47.13372)
3	Luxembourg	POINT (6.13000 49.61166)
4	Palikir	POINT (158.14997 6.91664)

Figure 4-7. *Head of the data. Image by author*
Data source: geopandas, BSD 3 Clause Licence

You can easily plot this dataset with the plot command, as we have done many times before. This is shown in Code Block 4-8.

Code Block 4-8. Plotting the cities data

```
cities.plot()
```

You will obtain a map with only points on it, as shown in Figure 4-8.

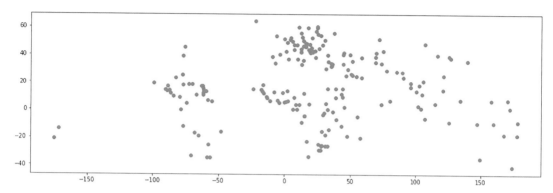

Figure 4-8. *Plot of the cities data. Image by author*
Data source: geopandas, BSD 3 Clause Licence

This plot is really not very readable. We need to add a background into this for more context. We can use the world's countries for this, using only the borders of the countries and leaving the content white.

The code in Code Block 4-9 does exactly that. It starts with creating the fig and ax and then sets the aspect to "equal" to make sure that the overlay will not be causing any mismatching. The world (country polygons) is then plotted using the color white to make it seem see-through, followed by the cities with a marker='x' for squares and the color='black' for black color.

Code Block 4-9. Adding a background to the cities data

```
import matplotlib.pyplot as plt
plt.rcParams["figure.figsize"] = [16,9]

fig, ax = plt.subplots()
ax.set_aspect('equal')
world.plot(ax=ax, color='white', edgecolor='grey')
cities.plot(ax=ax, marker='x', color='black', markersize=15)
plt.title('Cities plotted on a country border base map')
plt.show()
```

The resulting map looks as shown in Figure 4-9.

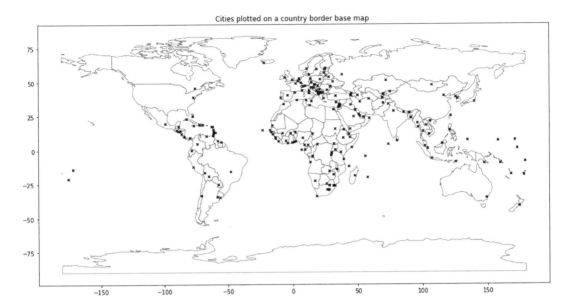

Figure 4-9. *Adding a background to the cities data. Image by author*
Data source: geopandas, BSD 3 Clause Licence

Concluding on Mapping with Geopandas and Matplotlib

This introductory example shows how to create basic maps with geopandas and matplotlib. Mapmaking and data visualization in general are creative processes, and although I can give you some pointers throughout this book, there is no one correct way to make visualizations.

If you want to go further into visualization and map mapping, I strongly recommend checking out the documentation of geopandas mapping and trying out different things to get yourself familiar with all the options that are out there.

Making a Map with Cartopy

We now move on to a second well-known map mapping library in Python called Cartopy. We'll be walking through one of the simpler examples of Cartopy that is shown in their documentation, in order to get a grasp of the different options, strong points, and weak points of this library.

A strong point of Cartopy is that they have a relatively extensive mapping gallery with maps that you can take and make your own. If you want to check out the examples on

their gallery, they can be found on this link: https://scitools.org.uk/cartopy/docs/
latest/gallery/lines_and_polygons/feature_creation.html#sphx-glr-gallery-
lines-and-polygons-feature-creation-py.

The example that we'll be looking at first is based on the example called "lines
and polygons feature creation." It allows you to make a basic land map of a region of
the earth.

The code in Code Block 4-10 shows you how this can be done. It goes through a
number of steps. The first important part of the code is to create the "fig.add_subplot"
in which you call a projection argument to solve any problems of coordinate systems
right from the start. In this case, the PlateCarree projection is chosen, but the cartopy.crs
library has a large number of alternatives.

Secondly, you see that ax.set_extent is called. This will make a subset of the map
ranging from the coordinate (-10, 30) in the left-bottom corner to (40, 70) in the top-right
corner. You'll see more details on setting extents in the next chapter.

After this, ax.stock_img() is called to add a background map. It is a simple but
recognizable background image with blue seas and brown/green land.

A number of built-in features from Cartopy are also added to the map: Land,
Coastline, and the states_provinces that come from Natural Earth Features and are set to
a 1:10 million scale.

Finally, a copyright is added to the bottom right.

Code Block 4-10. Creating a Cartopy plot

```
!pip install cartopy
import cartopy
import matplotlib.pyplot as plt
import cartopy.crs as ccrs
import cartopy.feature as cfeature
from matplotlib.offsetbox import AnchoredText

plt.rcParams["figure.figsize"] = [16,9]

fig = plt.figure()
ax = fig.add_subplot(1, 1, 1, projection=ccrs.PlateCarree())
ax.set_extent([-10, 40, 30, 70], crs=ccrs.PlateCarree())
```

```
# background image
ax.stock_img()

# use an inbuit feature from cartopy
states_provinces = cfeature.NaturalEarthFeature(
    category='cultural',
    name='admin_1_states_provinces_lines',
    scale='10m',
    facecolor='none')

ax.add_feature(cfeature.LAND)
ax.add_feature(cfeature.COASTLINE)
ax.add_feature(states_provinces, edgecolor='gray')

# Add a copyright
text = AnchoredText('\u00A9 Natural Earth; license: public domain',loc=4,
prop={'size': 12}, frameon=True)

ax.add_artist(text)

plt.show()
```

The map resulting from this introductory Cartopy example is shown in Figure 4-10.

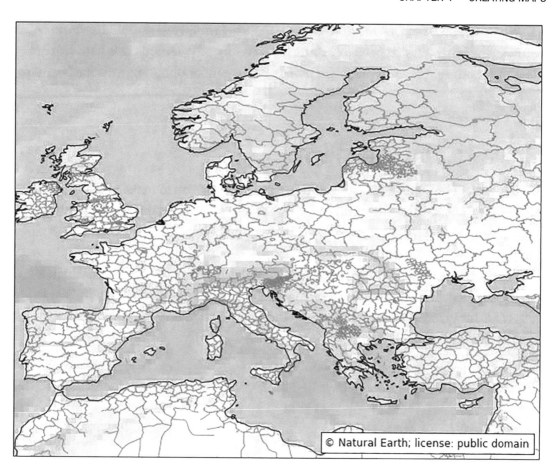

© Natural Earth; license: public domain

Figure 4-10. *The Cartopy map. Image by author*
Data source: Natural Earth, provided through Cartopy, Public Domain

Concluding on Mapping with Cartopy

Now that you have seen the example with Cartopy, it is good to make a comparison between the geopandas + matplotlib mapping technique and the Cartopy technique.

It seems that when working with geopandas, the focus is on the data. The resulting plot will look quite usable and has a great advantage due to its matplotlib syntax. Cartopy, on the other hand, is a library really made for mapping. It has no methods or best practices for storing data.

If you are really into the mapping features, you may prefer Cartopy, as it allows you to do a lot of things. On the other hand, it is a little less easy to manage, and you'll need to do more of the heavy lifting yourself than when you use geopandas.

When looking at the result in terms of aesthetics, I would argue that there is no clear winner. Similar results can probably be obtained with both methods. It is hard work to get an aesthetically pleasing map with both of these, but for obtaining informative maps, they work great.

Making a Map with Plotly

It is now time to move to a mapping library that will allow you to make much more aesthetically pleasing visualizations. One library that you can do this with is Plotly. Plotly is a well-known data visualization library in Python, and it has a well-developed component for mapping as well.

I strongly recommend checking out the Plotly map gallery (Figure 4-11), which is a great resource for getting started with mapmaking. Their plot gallery can be found at `https://plotly.com/python/maps/`, and it contains a large number of code examples of different types of graphs, including

- Choropleth maps

- Bubble maps

- Heat maps

- Scatter plots

- And much more

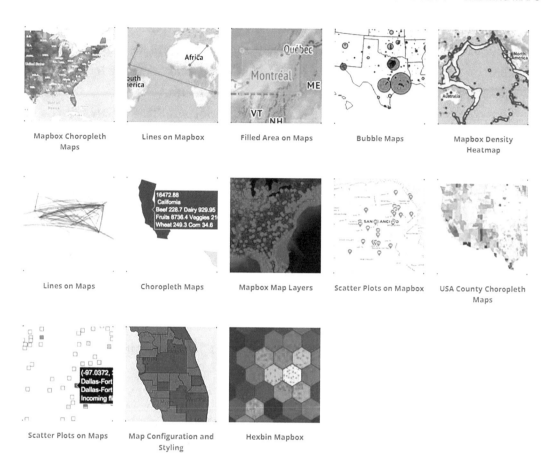

Mapbox Choropleth Maps	Lines on Mapbox	Filled Area on Maps	Bubble Maps	Mapbox Density Heatmap
Lines on Maps	Choropleth Maps	Mapbox Map Layers	Scatter Plots on Mapbox	USA County Choropleth Maps
Scatter Plots on Maps	Map Configuration and Styling	Hexbin Mapbox		

Figure 4-11. *Screenshot of the Plotly graph gallery (plotly.com/python/maps)*

To get a good grasp of the Plotly syntax, let's do a walk-through of a short Plotly example, based on their famous graphs in the graph gallery. In this example, you'll see how Plotly can easily make an aesthetically pleasing map, just by adding a few additional functionalities.

You can use the code in Code Block 4-11 to obtain a Plotly express example dataset that contains some data about cities.

Code Block 4-11. Map with Plotly

```
import plotly.express as px
data = px.data.gapminder().query("year==2002")
data.head()
```

The content of the first five lines of the dataframe tells us what type of variables we have. For example, you have life expectancy, population, and gdp per country per year. The filter on 2002 that was applied in the preceding query makes that we have only one data point per city; otherwise, plotting would be more difficult.

Let's create a new variable called gdp to make the plot with. This variable can be computed using Code Block 4-12.

Code Block 4-12. Adding a new variable gdpPerCap

```
data['gdp'] = data['gdpPercap'] * data['pop']
data.head()
```

Let's now make a bubble map in which the icon for each country is larger or smaller based on the newly created variable gdp using Code Block 4-13.

Code Block 4-13. Adding a variable gdp

```
fig = px.scatter_geo(data, locations="iso_alpha", size="gdp",
projection="natural earth")
fig.show()
```

Even with this fairly simple code, you'll obtain a quite interestingly looking graph, as shown in Figure 4-12.

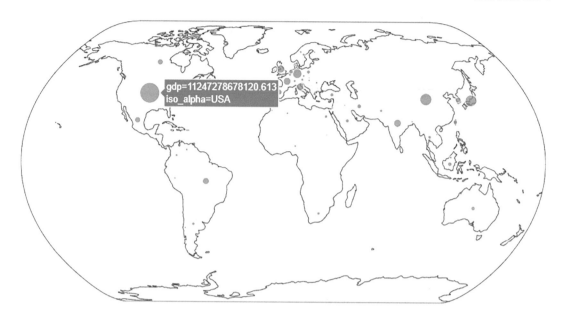

Figure 4-12. *A map with Plotly. Image by author*
Data source: Plotly data gapminder. Source data is free data from the World Bank via gapminder.org, CC-BY licence.

Each of the countries has a bubble that makes reference to their gdp, the continents each have a different color, and you have a background that is the well-known natural earth. You can hover over the data points to see more info about each of them.

Concluding on Mapping with Plotly

As you have probably concluded yourself, the real value of Plotly is in its simplicity. This library is great for making aesthetically pleasing maps without having to think long and hard about many of the complications that you may encounter. I invite you to look at more examples in Plotly to see other powerful examples and try out some of their galleries' example code.

Making a Map with Folium

In this fourth and final part on mapmaking with Python, you'll go another step further in the direction of better aesthetics. Folium is a Python library that integrates with a JavaScript library called Leaflet.js.

Folium allows you to create interactive maps, and the results will almost give you the feeling that you are working in Google Maps or comparable software. All this is obtained using a few lines of code, and all the complex work for generating those interactive maps is hidden behind Folium and Leaflet.js.

Folium has extensive documentation with loads of examples and quick-start tutorials (`https://python-visualization.github.io/folium/quickstart.html#Getting-Started`). To give you a feel for the type of results that can be obtained with Folium, we'll do a short walk-through of some great examples. Let's start by simply creating a map and then slowly adding specific parameters to create even better results.

Using the syntax in Code Block 4-14, you will automatically create a map of the Paris region. This is a one-line code, which just contains the coordinates of Paris, and it directly shows the power of the Folium library.

Code Block 4-14. Mapping with Folium

```
import folium
m = folium.Map(location=[48.8545, 2.2464])
m
```

You will obtain an interactive map in your notebook. It looks as shown in Figure 4-13.

```
import folium
m = folium.Map(location=[48.8545, 2.2464])
m
```

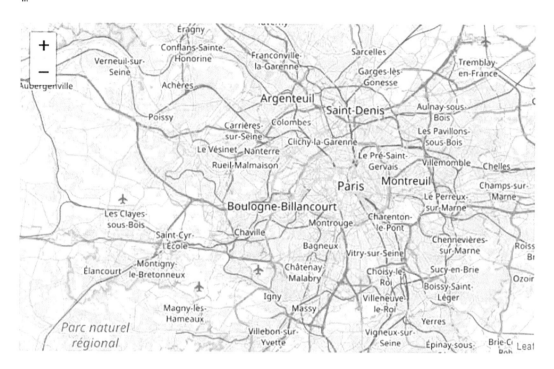

Figure 4-13. *A map with Folium. Image by author*
This Produced Work is based on the map data source from OpenStreetMap.
Map Data Source OpenStreetMap contributors. Underlying data is under Open
Database Licence. Map (Produced Work) is under the copyright of this book.
`https://wiki.osmfoundation.org/wiki/Licence/Licence_and_Legal_`
`FAQ#What_is_the_licence,_how_can_I_use_it?`

Now, the interesting thing here is that this map does not contain any of your data. It seems like it could be a map filled with complex points, polygons, labels, and more, and deep down somewhere in the software it is. The strong point of Folium as a visualization layer is that you do not at all need to worry about this. All your "background data" will stay cleanly hidden from the user. You can imagine that this would be very complex to create using the actual polygons, lines, and points about the Paris region.

Let's go a step further and add some data to this basemap. We'll add two markers (point data in Folium terminology): one for the Eiffel Tower and one for the Arc de Triomphe.

The code in Code Block 4-15 shows a number of additions to the previous code. First, it adds a zoom_start. This basically tells you how much zoom you want to show when initializing the map. If you have played around with the first example, you'll see that you can zoom out so far as to see the whole world on your map and that you can zoom in to see a very detailed map as well. It really is very complete. However, for a specific use case, you would probably want to focus on a specific region or zone, and setting a zoom_start will help your users identify what they need to look at.

Second, there are two markers added to the map. They are very intuitively added to the map using the .add_to method. Once added to the map, you simply show the map like before, and they will appear. You can specify a popup so that you see additional information when hovering over your markers. Using HTML markup, you can create whole paragraphs of information here, in case you'd want to.

As the markers are point geometry data, they just need x and y coordinates to be located on the map. Of course, these coordinates have to be in the correct coordinate system, but that is nothing different from anything you've seen before.

Code Block 4-15. Add items to the Folium map

```
import folium
m = folium.Map(location=[48.8545, 2.2464], zoom_start=11)

folium.Marker(
    [48.8584, 2.2945], popup="Eiffel Tower").add_to(m)
folium.Marker(
    [48.8738, 2.2950], popup="Arc de Triomphe").add_to(m)

m
```

If you are working in a notebook, you will then be able to see the interactive map appear as shown in Figure 4-14. It has the two markers for showing the Eiffel Tower and the Arc de Triomphe, just like we started out to do.

For more details on plotting maps with Folium, I strongly recommend you to read the documentation. There is much more documentation out there, as well as sample maps and examples with different data types.

```
import folium
m = folium.Map(location=[48.8545, 2.2464], zoom_start=11)

folium.Marker(
    [48.8584, 2.2945], popup="Eiffel Tower").add_to(m)
folium.Marker(
    [48.8738, 2.2950], popup="Arc de Triomphe").add_to(m)

m
```

Figure 4-14. *Improved Folium map. Image by author*
This Produced Work is based on the map data source from OpenStreetMap.
Map Data Source OpenStreetMap contributors. Underlying data is under Open
Database Licence. Map (Produced Work) is under the copyright of this book.
`https://wiki.osmfoundation.org/wiki/Licence/Licence_and_Legal_`
`FAQ#What_is_the_licence,_how_can_I_use_it?`

Concluding on Mapping with Folium

If making interactive, user-friendly, aesthetically pleasing maps is what you're looking for, you'll discover a great tool in Folium. Its built-in background maps can be a strong advantage if you're looking to plot little datasets in such a way to obtain a geospatial context through the background maps, without having to obtain any data about this background data.

Folium allows you to create beautiful interactive maps with little data and is therefore great for visualization-oriented use cases. For scientific maps and more analytical and mathematical purposes, a disadvantage could be that the background maps take the eyes off of what your analysis is trying to showcase. Less fancy maps could be a better choice in that case.

Key Takeaways

1. There are many mapping libraries in Python, each with its specific advantages and disadvantages.

2. Using geopandas together with matplotlib is probably the easiest and most intuitive approach to making maps with Python. This approach allows you to work with your dataframes in an intuitive pandas-like manner in geopandas and use the familiar matplotlib plotting syntax. Aesthetically pleasing maps may be a little bit of work to obtain.

3. Cartopy is an alternative that is less focused on data and more on the actual mapping part. It is a very specific library to cartography and has good support for different geometries, different coordinate systems, and the like.

4. Plotly is a visualization library, and it is, therefore, less focused on the geospatial functionalities. It does come with a powerful list of visualization options, and it can create aesthetically pleasing maps that can really communicate a message.

5. Folium is a great library for creating interactive maps. The maps that you can create even with little code are of high quality and are similar in user experience to Google Maps and the like. The built-in background maps allow you to make useful maps even when you have very little data to show.

6. Having seen those multiple approaches to mapmaking, the most important takeaway is that maps are created for a purpose. They either try to make an objective representation of some data, or they can try to send a message. They can also be made for having something that is nice to look at. When choosing your method for making maps with Python, you should choose the library and the method that best serves your purpose. This always depends on your use case.

PART II

GIS Operations

CHAPTER 5

Clipping and Intersecting

In the previous four chapters, you have discovered the foundations of working with geodata. In the first chapter, you have seen what geodata is, how to represent it, and a general overview of tools for using geodata. After that, Chapter 2 has given you a deeper introduction to coordinate systems and projections, which gives you a framework in which coordinates can be used.

Chapter 3 has shown you how to represent geographical data in practice, using one of the geographical data types or shapes. Chapter 4 has given you an introduction to creating maps.

At this point, you should start having a general idea of geodata and start getting some understanding of what can be done with it, both in a general, theoretical sense and in a practical sense through the Python examples that have been presented.

Now that this introduction is done, the remainder of this book will serve two main objectives. In the coming four chapters, we will focus on four standard operations in geodata processing. The goal here will be to show how to use Python for tasks that are generally already implemented in more specific GIS tools and systems. When using Python for your GIS work, it is essential that you have these types of standard geodata processing tasks in your personal toolkit.

More advanced use cases using machine learning are much less commonly implemented in GIS tools, and their easy availability is what makes Python a great choice as a tool for geodata. These use cases will be covered in the final chapters of the book.

The standard operations that will be covered are

- Clipping and intersecting

- Buffering

- Merge and dissolve

- Erase

© Joos Korstanje 2022
J. Korstanje, *Machine Learning on Geographical Data Using Python*,
https://doi.org/10.1007/978-1-4842-8287-8_5

The first of the standard operations, clipping and intersecting, is covered in this chapter. Let's start by giving a general definition of the clipping operation and do an example in Python. We will then do the same for the intersecting operation.

What Is Clipping?

Clipping, in geoprocessing, takes one layer, an input layer, and uses a specified boundary layer to cut out a part of the input layer. The part that is cut out is retained for future use, and the rest is generally discarded.

The clipping operation is like a cookie cutter, in which your cookie dough is the input layer in which a cookie-shaped part is being cut out.

A Schematic Example of Clipping

Let's clarify this definition using a more intuitive example. Imagine that you are working with geodata of a public park. You can get a lot of public data from the Internet, but of course any public data is unlikely to be exactly catered to the size of your park: it will generally be on a much larger scale.

To make working with the data easier, you can use a clipping operation to get all the data back to a smaller size and keep only data that is relevant for your park. The schematic drawing in Figure 5-1 shows how this would work.

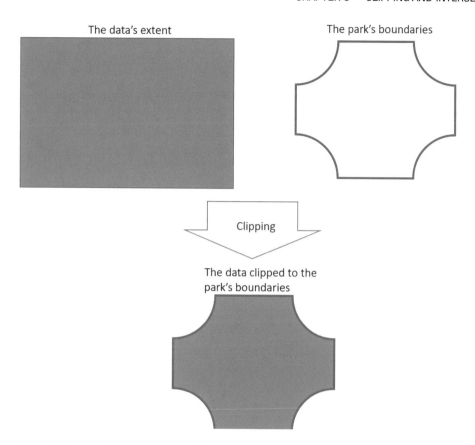

Figure 5-1. *Schematic overview of the clipping operation. Image by author*

What Happens in Practice When Clipping?

As you have understood by now, any mapping information is just data. Let's try to understand what happens to the data when we are executing a clipping operation.

In practice, the clipping operation can have multiple effects, depending on the input data. If you are working with raster data, a clipping operation would simply remove the pixels of the raster that are not selected and keep the pixels that are still relevant. The previous schematic drawing shows what would happen with raster data. You can consider that the input data consists of a large number of pixels. In the output, the nonrelevant pixels have been deleted.

With point data, the same is true: some points will be selected, other points not. In terms of data, this means dropping the rows of data that are not to be retained. The schematic drawing in Figure 5-2 shows this case.

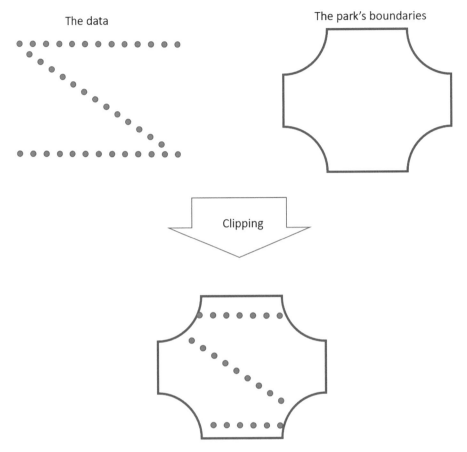

Figure 5-2. *Schematic drawing of clipping points. Image by author*

When clipping a line dataset, things become more complicated, as lines may start inside the clip boundaries and end outside of them. In this case, the part of the line that is inside the boundary has to be kept, but the part that is outside of the boundary has to be removed. The result is that some rows of data will be entirely removed (lines that are completely out of scope) and some of them will be altered (lines that are partly out of scope). This will be clearer with a more detailed schematic drawing that is shown in Figure 5-3. In this schematic drawing, the data is line data; imagine, for example, a road network. The clip is a polygon.

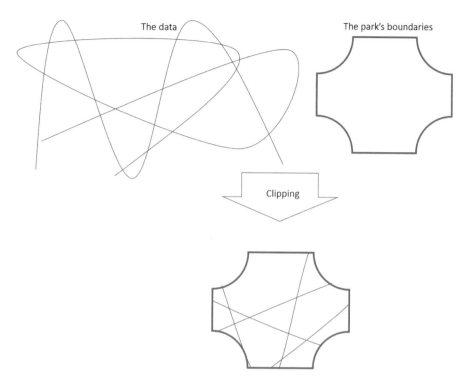

Figure 5-3. *Schematic drawing of clipping lines. Image by author*

In the coming part, you will see a more practical application of this theory by applying the clipping operation in Python.

Clipping in Python

In this example, you will see how to apply a clipping operation in Python. The dataset is a dataset that I have generated specifically for this exercise. It contains two features:

- A line that covers a part of the Seine River (a famous river in Paris, France, which also covers a large part of the country of France)

- A polygon that covers the center of Paris

The goal of the exercise is to clip the Seine River to the Paris center region. This is a very realistic use of the clipping operation. After all, rivers are often multicountry objects and are often displayed in maps. When working on a more local map, you will likely encounter the case where you will have to clip rivers (or other lines like highways, train lines, etc.) to a more local extent.

Let's start with importing the dataset and opening it. You can find the data in the GitHub repository. For the execution of this code, I'd recommend using a Kaggle notebook or a local environment, as Colab has an issue with the clipping function at the time of writing.

You can import the data using geopandas, as you have learned in previous chapters. The code for doing this is shown in Code Block 5-1.

Code Block 5-1. Importing the data

```
import geopandas as gpd
import fiona

gpd.io.file.fiona.drvsupport.supported_drivers['KML'] = 'rw'
data = gpd.read_file('ParisSeineData.kml')
print(data)
```

The data looks as shown in Figure 5-4.

```
   Name Description                                            geometry
0  Seine              LINESTRING Z (2.11750 48.91626 0.00000, 2.1106...
1  Paris              POLYGON Z ((2.25631 48.84059 0.00000, 2.35622 ...
```

Figure 5-4. *The dataset. Image by author*

We can quickly use the geopandas built-in plot function to get a plot of this data. Of course, you have already seen more advanced mapping options in the previous chapters, but the goal here is just to get a quick feel of the data we have. This is done in Code Block 5-2.

Code Block 5-2. Plotting the data

```
data.plot()
```

When using this plot method, you will observe the map in Figure 5-5, which clearly contains the two features: the Seine River as a line and the Paris center as a polygon.

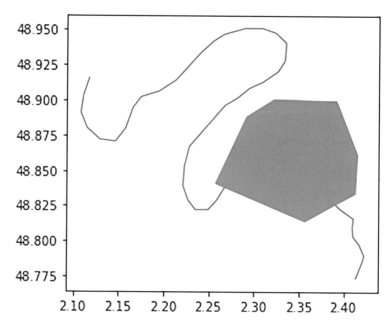

Figure 5-5. *The plot resulting from Code Block 5-2. Image by author*

Now, as stated in the introduction of this example, the goal is to have only the Seine River line object, but to clip it to the size of the Paris river. The first step is to split our data object into two separate objects. This way, we will have one geodataframe with the Seine River and a second geodataframe with the Paris polygon. This will be easier to work with. You can extract the Seine River using the code in Code Block 5-3.

Code Block 5-3. Extract the Seine data

```
seine = data.iloc[0:1,:]
seine.plot()
```

You can verify in the resulting plot (Figure 5-6) that this has been successful.

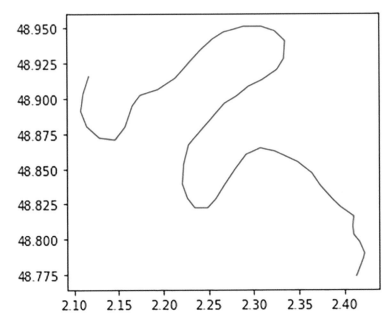

Figure 5-6. *The plot resulting from Code Block 5-3. Image by author*

Now, we do the same for the Paris polygon using the code in Code Block 5-4.

Code Block 5-4. Extracting the Paris data

```
paris = data.iloc[1:2,:]
paris.plot()
```

You will obtain a plot with the Paris polygon to verify that everything went well. This is shown in Figure 5-7.

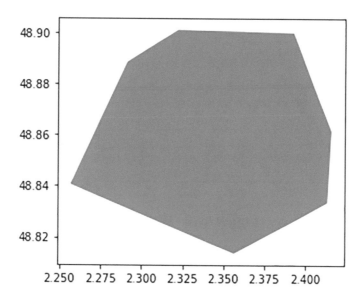

Figure 5-7. *The plot resulting from Code Block 5-4. Image by author*

Now comes the more interesting part: using the Paris polygon as a clip to the Seine River. The code to do this using geopandas is shown in Code Block 5-5.

Code Block 5-5. Clipping the Seine to the Paris region

```
paris_seine = seine.clip(paris)
paris_seine
```

You will obtain a new version of the Seine dataset, as shown in Figure 5-8.

	Name	Description	geometry
0	Seine		LINESTRING Z (2.26703 48.83771 0.00000, 2.2678...

Figure 5-8. *The dataset after clipping. Image by author*

You can use the code in Code Block 5-6 to plot this version to see that it contains only those parts of the Seine River that are inside the Paris center region.

Code Block 5-6. Plotting the clipped data

```
paris_seine.plot()
```

You will see the result in Figure 5-9.

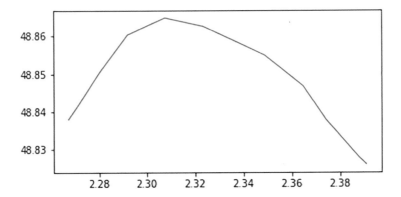

Figure 5-9. *The Seine River clipped to the Paris polygon. Image by author*

This result shows that the goal of the exercise is met. We have successfully imported the Seine River and Paris polygon, and we have reduced the size of the Seine River line data to fit inside Paris.

You can imagine that this can be applied for highways, train lines, other rivers, and other line data that you'd want to use in a map for Paris, but that is available only for a much larger extent. The clipping operation is fairly simple but very useful for this, and it allows you to remove useless data from your working environment.

What Is Intersecting?

The second operation that we will be looking at is the intersection. For those of you who are aware of set theory, this part will be relatively straightforward. For those who are not, let's do an introduction of set theory first.

Sets, in mathematics, are collections of unique objects. A number of standard operations are defined for sets, and this is generally helpful in very different problems, one of which is geodata problems.

As an example, we could imagine two sets, A and B:

- Set A contains three cities: New York, Las Vegas, and Mexico City.

- Set B contains three cities as well: Amsterdam, New York, and Paris.

There are a number of standard operations that are generally applied to sets:

- Union: All elements of both sets

- Intersection: Elements that are in both sets

- Difference: Elements that are in one but not in the other (not symmetrical)

- Symmetric difference: Elements that are in A but not in B or in B but not in A

With the example sets given earlier, we would observe the following:

- The union of A and B: New York, Las Vegas, Mexico City, Amsterdam, Paris

- The intersection of A and B: New York

- The difference of A with B: Las Vegas, Mexico City

- The difference of B with A: Amsterdam, Paris

- The symmetric difference: Las Vegas, Mexico City, Amsterdam, Paris

The diagram in Figure 5-10 shows how these come about.

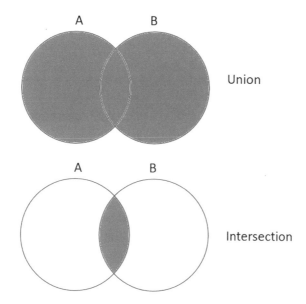

Figure 5-10. *part 1: Set operations. Image by author*

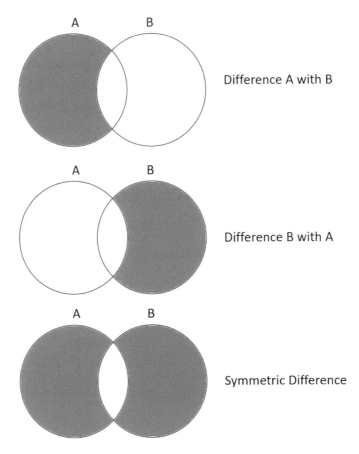

Figure 5-10. *part 2: More set operations. Image by author*

In the first part of this chapter, you have seen that filtering is an important basic operation in geodata. Set theory is useful for geodata, as it allows you to have a common language for all these filter operations.

The reason that we are presenting the intersection in the same chapter as the clip is that they are relatively similar and are often confused. This will allow us to see what the exact similarities and differences are.

What Happens in Practice When Intersecting?

An intersection in set theory takes two input sets and keeps only those items from the set that are present in both. In geodata processing, the same is true. Consider that your sets are now geographical datasets, in which we use the geographical location data as identifier of the objects. The intersection of two objects will keep all features (columns) of both datasets, but it will keep only those data points that are present in both datasets.

As an example, let's consider that we again use the Seine River data, and this time we use the main road around Paris (Boulevard Périphérique) to identify places at which we should find bridges or tunnels. This could be useful, for example, if we have no data about bridges and tunnels yet, and we want to automatically identify all locations at which we should find bridges or tunnels.

The intersection of the two would allow us to keep both the information about the road data and the data from the river dataset while reducing the data to the locations where intersections are to be found.

Of course, this can be generalized to a large number of problems where the intersection of two datasets is needed.

Conceptual Examples of Intersecting Geodata

Let's now see some examples of how the intersection operation can apply to geodata. Keep in mind that in raster data, there are no items, just pixels, so using set theory is only useful for vector data. You will now see a number of examples of the different vector data types and how an intersection would work for them.

Let's start with considering what happens when intersecting points with points. As a point has no width nor length, only a location, the only intersection that we can do is identify whether points from one dataset overlap with points from the other dataset. If they do, we can consider that these overlapping points are in the intersection. The schematic drawing in Figure 5-11 shows an example of this.

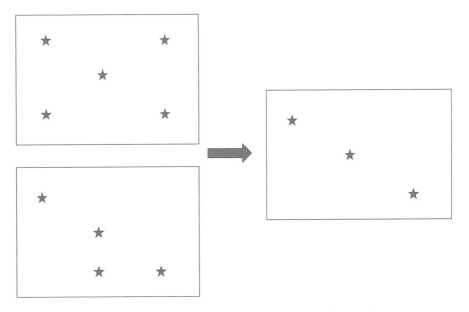

Figure 5-11. *Schematic drawing of intersecting. Image by author*

This basically just filters out some points, and the resulting shapes are still points. Let's now see what happens when applying this to two line datasets.

Line datasets will work differently. When two lines have a part at the exact same location, the resulting intersection of two lines could be a line. In general, it is more likely that two lines intersect at a crossing or that they are touching at some point. In this case, the intersection of two lines is a point. The result is therefore generally a different shape than the input. This is shown in the schematic drawing in Figure 5-12.

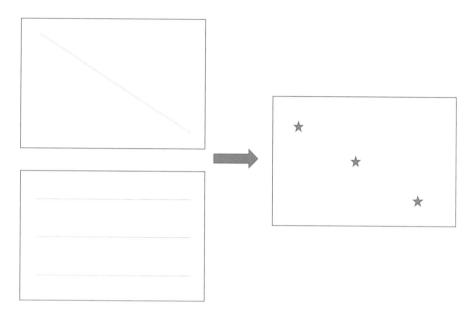

Figure 5-12. *Schematic drawing of intersecting lines. Image by author*

The lines intersect at three points, and the resulting dataset just shows these three points. Let's now see what happens when intersecting polygons.

Conceptually, as polygons have a surface, we consider that the intersection of two polygons is the surface that they have in common. The result would therefore be the surface that they share, which is a surface and therefore needs to be polygon as well. The schematic drawing in Figure 5-13 shows how this works.

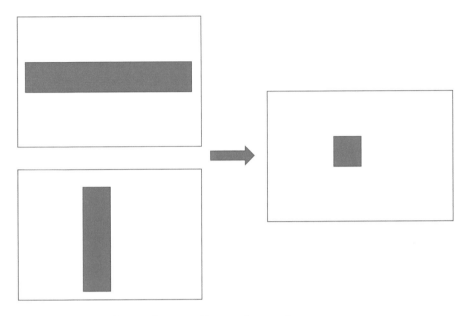

Figure 5-13. *Intersecting polygons. Image by author*

The result is basically just one or multiple smaller polygons. In the following section, you will see how to apply this in Python.

Intersecting in Python

Let's now start working on the example that was described earlier in this chapter. We take a dataset with the Boulevard Périphérique and the Seine River, and we use the intersection of those two to identify the locations where the Seine River crosses the Boulevard Périphérique.

You can use the code in Code Block 5-7 to import the data and print the dataset.

Code Block 5-7. Import and print the data

```
gpd.io.file.fiona.drvsupport.supported_drivers['KML'] = 'rw'
data = gpd.read_file('ParisSeineData_example2_v2.kml')
data.head()
```

You will observe the data in Figure 5-14.

	Name	Description	geometry
0	Boulevard Peripherique		POLYGON Z ((2.25190 48.84116 0.00000, 2.25155 ...
1	Seine		POLYGON Z ((2.21962 48.83890 0.00000, 2.22546 ...

Figure 5-14. *The data. Image by author*

There are two polygons, one called Seine and one called Boulevard Périphérique. Let's use Code Block 5-8 to create a plot to see what this data practically looks like. We can use the cmap to specify a colormap and obtain different colors. You can check out the matplotlib documentation for an overview of colormaps; there are many to choose from.

Code Block 5-8. Plot the data with a colormap

```
data.plot(cmap='tab10')
```

You will obtain the result in Figure 5-15.

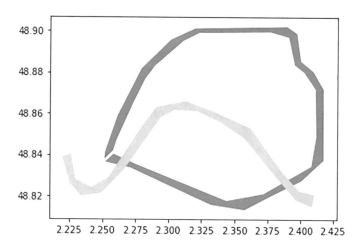

Figure 5-15. *The plot resulting from Code Block 5-8. Image by author*

Compared to the previous example, the data has been converted to polygons here. You will see in a later chapter how to do this automatically using buffering, but for now it has been done for you, and the polygon data is directly available in the dataset.

We can clearly see two intersections, so we can expect two bridges (or tunnels) to be identified. Let's now use the intersection function to find these automatically for us.

The code in Code Block 5-9 shows how to use the overlay function in geopandas to create an intersection.

Code Block 5-9. Creating an intersection

```
# Extract periph data
periph = data.iloc[0:1,:]

# Extract seine data
seine = data.iloc[1:2,:]

intersection = seine.overlay(periph, how='intersection')
intersection
```

The result is a dataset with only the intersection of the two polygons, as shown in Figure 5-16.

	Name_1	Description_1	Name_2	Description_2	geometry
0	Boulevard Peripherique		Seine		MULTIPOLYGON Z (((2.39094 48.82825 0.00000, 2....

Figure 5-16. *The plot resulting from Code Block 5-9. Image by author*

The resulting object is a multipolygon, as it contains two polygons: one for each bridge (or tunnel). You can see this more easily when creating the plot of this dataset using Code Block 5-10.

Code Block 5-10. Plotting the intersection

```
intersection.plot()
```

The result may look a bit weird without context, but it basically just shows the two bridges/tunnels of the Parisian Boulevard Périphérique. This is shown in Figure 5-17.

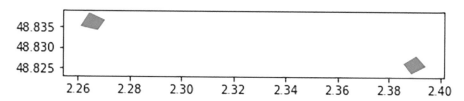

Figure 5-17.   *The two crossings of the Seine and the Boulevard Périphérique. Image by author*

The goal of the exercise is now achieved. We have successfully created an automated method for extracting locations where roads cross rivers. If we would now want to do this for the whole city, we could simply find datasets with all Paris's roads and rivers and use the same method to find all the bridges in Paris.

Of course, this was just one example, and you can generalize this to many situations where creating intersections is useful, for example, to create new features in your dataset if you want to do machine learning or for adding new features on maps.

It will be useful to keep in mind that there are other options in the overlay function, of which we will see some in coming chapters. These are all related to other operations in set theory, which is a very practical way to think about these basic geodata operations.

Difference Between Clipping and Intersecting

Now that you have seen both the clipping and the intersection tools, you should understand that both of them generally reduce the quantity of data. There is a fundamental difference between the two, and your choice for the tool should depend on what you intend to accomplish.

Clipping reduces data by taking an input dataset and a boundary dataset. The resulting output data is of the exact shape of the input data, only limited to the geographic extent imposed by the boundary dataset. The only data that is kept is the data of the input layer. You have seen in the clipping example that the Paris polygon was not present in the output: only the Seine River was present, yet in a geographically reduced form.

With intersections, both datasets can be considered as equally important. There is not one input dataset and one boundary dataset, but there are two input datasets. The resulting output is a combination that keeps all the information from both input datasets while still keeping only points that coincide geographically.

In the intersection example that will be done after this, you'll see that the output contains data from both input datasets and that it is therefore different from the clipping operation.

Key Takeaways

1. There are numerous basic geodata operations that are standardly implemented in most geodata tools. They may seem simple at first sight, but applying them to geodata can come with some difficulties.

2. The clipping operation takes an input dataset and reduces its size to an extent given by a boundary dataset. This can be done for all geodata data types.

3. Using clipping for raster data or points comes down to deleting the pixel points that are out of scope.

4. Using clipping for lines or polygons will delete those lines and polygons that are out of scope entirely, but will create a new reduced form for those points that are partly inside and partly outside of the boundaries.

5. The intersection operation is based on set theory and allows to find features that are shared between two input datasets. It is different from clipping, as it treats the two datasets as input and therefore keeps the features of both of them. In clipping, this is not the case, as only the features from the input dataset are considered relevant.

6. Intersecting points basically comes down to filtering points based on their presence in both datasets.

7. Intersecting lines generally results in points (either crossings between two lines or touchpoints between two curving lines), but they can also be lines if two lines are perfectly equal on a part of their trajectory.

8. Intersecting polygons will result in one or multiple smaller polygons, as the intersection is considered to be the area that the two polygons have in common.

9. You have seen how to use geopandas as an easy tool for both clipping and intersecting operations.

CHAPTER 6

Buffers

In the previous chapter, we have started looking at a number of common geospatial operations: data operations that are not possible, or at least not common, on regular data, but that are very common on geospatial data.

The standard operations that will be covered are

- Clipping and intersecting

- Buffering

- Merge and dissolve

- Erase

In the previous chapter, you have already seen two major operations. You have first seen how to clip data to a specific extent, mainly for the use of dropping data based on a spatial range. You have also seen how to use intersecting to create data based on applying set theory on geospatial datasets. It was mentioned that other set theory operations can be found in that scope as well.

In this chapter, we will look at a very different geospatial operation. You will discover the geospatial operation of buffering or creating buffers. They are among the standard operations of geospatial operations, and it is useful to master this tool.

Just like intersecting, the buffer is a tool that can be used either as a stand-alone or as a tool for further analysis. It was not mentioned, but in the example of intersections, a buffer operation was used to create polygon data of the bridges and rivers.

This clearly shows how those spatial operations should all be seen as tools in a toolkit, and when you want to achieve a specific goal, you need to select the different tools that you need to get there. This often means using different combinations of tools in an intelligent manner. The more tools you know, the more you will be able to achieve.

© Joos Korstanje 2022
J. Korstanje, *Machine Learning on Geographical Data Using Python*,
https://doi.org/10.1007/978-1-4842-8287-8_6

We will start this chapter with an introduction into the theory behind the buffer, and then do a number of examples in Python, in which you will see how to create buffers using different geodata types.

What Are Buffers?

Buffers are newly created polygons that surround other existing shapes. The term buffering is adequate, as it really consists of creating a buffer around an existing object.

You can create buffers around all vector data objects, and they always become polygons. For example, this would give the following results based on data type:

- When you create a buffer around a point, you will end up with a new buffer polygon that contains the surrounding area around that point.

- By adding a buffer to a line, you have a new feature that contains the area around that line.

- Buffers around polygons will also contain the area just outside the polygon.

Buffers are newly created objects. After computing the buffer polygon, you still have your original data. You will simply have created a new feature which is the buffer.

A Schematic Example of Buffering

Let's clarify this definition using a more intuitive example. We will take the example that was used in the previous chapter as well, in which we had a line dataset of roads, but we needed to transform those roads into polygons to take into account the width of the road. In the image in Figure 6-1, you see a road depicted first as a polygon and then as a line.

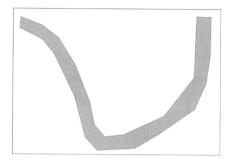 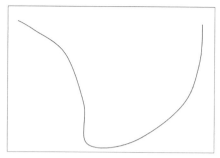

Figure 6-1. Showing the road first as a polygon and then as a line. Image by author

Representing the road as a line will allow you to do many things, like compute the length of a road, find crossings with other roads and make a road network, etc. However, in real life, a road has a width as well. For things like urban planning around the road, building a bridge, etc., you will always need to know the width of the road at each location.

As you have seen when covering data types in Chapter 3, lines have a length but not width. It would not be possible to represent the width of a line. You could however create a buffer around the line and give the buffer a specified width. This would result in a polygon that encompasses the road, and you would then be able to generate a polygon-like data.

What Happens in Practice When Buffering?

Let's now see some examples of how the intersection operation can apply to geodata. Keep in mind that in raster data, there are no items, just pixels, so using set theory is only useful for vector data. You will now see a number of examples of the different vector data types and how an intersection would work for them.

Buffers for Point Data

Let's start with considering what happens when constructing buffers around points. Points have no width or length, just location. If we create a buffer, we generally make buffers of which the borders are equally far away from the buffered object. Therefore, buffers around points would generally be circles. The chosen size of the buffer would determine the size of those circles. You can see an example in the schematic drawing in Figure 6-2.

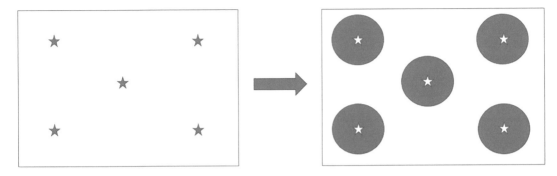

Figure 6-2. *Buffers for point data. Image by author*

In this schematic drawing, you see the left image containing points, which are depicted here as stars. In the right image, you see how the buffers are circular polygons that are formed exactly around the point.

Although it may seem difficult to find use cases for this, there are cases where this may be useful. Imagine that your point data are sources of sound pollution and that it is known that the sound can be heard a given number of meters from the source. Creating buffers around the point would help to determine regions in which the sound problems occur.

Another, very different use case could be where you collect data points that are not very reliable. Imagine, for example, that they are gps locations given by a mobile phone. If you know how much uncertainty there is in your data points, you could create buffers around your data points that state that all locations that are inside the buffer may have been visited by the specific mobile phone user. This can be useful for marketing or ad recommendations and the like.

Buffers for Line Data

Let's now consider the case in which we apply buffers to line data. A first schematic example was already given before. In that example, a line dataset was present, but it needed to be converted into a polygon to use it for the intersection exercise.

Let's consider another example. Imagine that you are planning a railroad location somewhere, and you need to investigate the amount of noise problems that you are going to observe and how many people this is going to impact. This is an interesting example, as it advances on the sound pollution problem introduced earlier.

We could imagine building different buffers around the railroad (one for very strongly impacted locations and one for moderately impacted houses). The schematic drawing in Figure 6-3 shows how building different buffers around the railroad could help you in solving this problem.

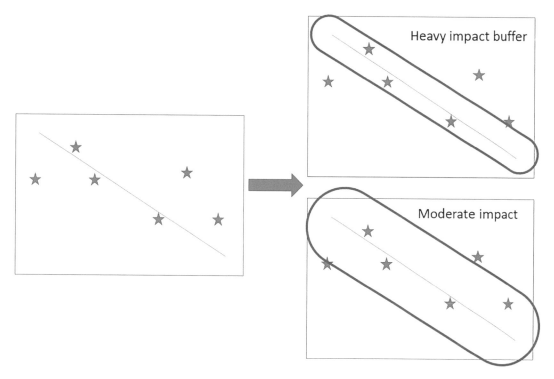

Figure 6-3. *Buffers for line data. Image by author*

You see that the left image contains one line (the planned railroad) and a number of houses (depicted as stars). On the top right, you see a narrow buffer around the line, which shows the heavy impact. You could filter out the points that are inside this heavy impact buffer to identify them in more detail. The bottom-left graph contains houses with a moderate impact. You could think of using set operations from the previous chapter to select all moderate impact houses that are not inside the heavy impact buffer (e.g., using a difference operation on the buffer, but other approaches are possible as well).

Buffers for Polygon Data

The third and last data type that we'll discuss for buffering is the polygon. Imagine that you have a geographical dataset of an area of a city, in which you have a lake as a feature. This lake is well known to have a path around it, yet the path is not visible on the map. You can decide to use a buffer around the lake to make a quick estimation of the path's location. The schematic drawing in Figure 6-4 shows the idea.

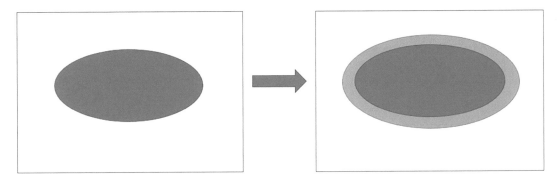

Figure 6-4. *Buffers for polygon data. Image by author*

In the left part of the schematic drawing, you see the lake polygon, an oval. On the right, you see that a gray buffer has been created around the lake – maybe not the best way to estimate the exact location of your path, but definitely an easy way to create the new feature quickly in your dataset.

Now that you have seen how buffers work in theory, it is time to move on to some practice. In the following section, we will start applying these operations in Python.

Creating Buffers in Python

In the coming example, we will be looking at a quite simple but intuitive example of searching a house to rent. Imagine that you have a database with geographic point data that are houses, and you want to do a geodata lookup to test for a number of criteria. In this example, we will go through a number of examples using buffers. You will discover how you can use buffers for your house searching criteria:

- You will see how to use buffers around points to investigate whether each of the houses is on walking distance to a subway station (a real added value).

- You will then see how to use buffers around a subway line to make sure that the house is not disturbed by the noises of the subway line (which can be a real problem).

- You will then see how to use buffers around parks to see whether each of the houses is at least in walking distance of a park.

Let's start with the first criterion by creating buffers around some subway stations.

Creating Buffers Around Points in Python

Let's start by importing the .kml data. Having gone through this numerous times in early chapters, this should start to become a habit by now. The code to import the data is shown in Code Block 6-1.

Code Block 6-1. Import the data

```
import geopandas as gpd
import fiona

# import the paris subway station data
gpd.io.file.fiona.drvsupport.supported_drivers['KML'] = 'rw'
data = gpd.read_file('Paris_Metro_Stations.kml')
data
```

The resulting dataframe is shown in Figure 6-5.

	Name	Description	geometry
0	Station 1		POINT Z (2.37447 48.84432 0.00000)
1	Station 2		POINT Z (2.39695 48.84886 0.00000)
2	Station 3		POINT Z (2.43332 48.84737 0.00000)
3	Station 4		POINT Z (2.48693 48.85342 0.00000)
4	Station 5		POINT Z (2.34674 48.86190 0.00000)
5	Station 6		POINT Z (2.32936 48.87226 0.00000)
6	Station 7		POINT Z (2.29493 48.87381 0.00000)
7	Station 8		POINT Z (2.23821 48.89170 0.00000)

Figure 6-5. *The point data. Image by author*

You will see that the data contains eight subway stations. They do not have names as that does not really have added value for this example. They are all point data, having a latitude and longitude. They also have a z-score (height), but they are not used and they are therefore all at zero.

Let's make a quick and easy visualization to get a better feeling for the data that we are working with. You can use the code in Code Block 6-2 to do so.

Code Block 6-2. Plotting the data

```
data.plot()
```

This plot will show the plot of the data. This is shown in Figure 6-6.

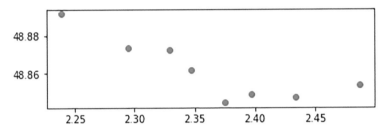

Figure 6-6. *The plot resulting from Code Block 6-2. Image by author*

We can clearly imagine the points being stations of a subway line. This plot is not very visual. If you want to work on visuals, feel free to add some code from Chapter 4 to create background visuals. You can also use the contextily library, which is a Python package that can very easily create background maps. The code in Code Block 6-3 shows how it is done. It uses the example data from Chapter 5 to create a background map with a larger extent.

Code Block 6-3. Make a plot with a background map

```
# !pip install contextily
import contextily as cx

# import the paris subway station data
gpd.io.file.fiona.drvsupport.supported_drivers['KML'] = 'rw'
paris = gpd.read_file('ParisSeineData.kml')
paris = paris.loc[1:,:]

# use paris data to set extent but leave invisible
ax = paris.plot(figsize=(15,15), color="None")
```

```
# add the data
data.plot(ax=ax)
```

```
# add the background map
cx.add_basemap(ax, crs=data.crs)
```

This will result in the map shown in Figure 6-7.

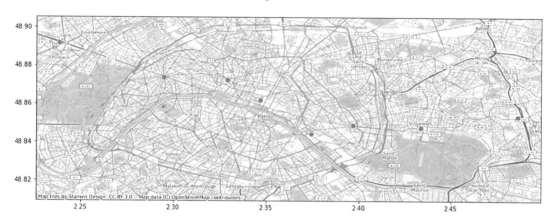

Figure 6-7. *The map with a background. Image by author using contextily source data and image as referenced in the image*

As you can see, the points are displayed on the map, on the subway line that goes east-west. When we add houses to this data, we could compute distances from each house to each subway station. However, we could not use these points in a set operation or overlay. The overlay method would be much easier to compute than the distance operation, which shows why it is useful to master the buffer operation.

We can use it to combine with other features as specified in the definition of the example. Let's now add a buffer on those points to start creating a house selection polygon.

Creating the buffer is quite easy. It is enough to use ".buffer" and specify the width, as is done in Code Block 6-4.

Code Block 6-4. Creating the buffer

```
data.buffer(0.01)
```

This buffer operation will generate a dataset in polygons that now contains all the buffer polygons, as shown in Figure 6-8.

```
0    POLYGON ((2.38447 48.84432, 2.38442 48.84334, ...
1    POLYGON ((2.40695 48.84886, 2.40690 48.84788, ...
2    POLYGON ((2.44332 48.84737, 2.44327 48.84639, ...
3    POLYGON ((2.49693 48.85342, 2.49688 48.85244, ...
4    POLYGON ((2.35674 48.86190, 2.35669 48.86092, ...
5    POLYGON ((2.33936 48.87226, 2.33931 48.87128, ...
6    POLYGON ((2.30493 48.87381, 2.30488 48.87283, ...
7    POLYGON ((2.24821 48.89170, 2.24816 48.89072, ...
```

Figure 6-8. *The buffer polygons. Image by author*

It is theoretically better to do this only with Projected Coordinate Systems, which is not our case here, but for this example it is really not that impacting. Let's do without the conversions to keep this example easier to follow.

Now, let's add this buffer to our plot, to be able to visualize the areas in which we could select houses that meet the criterion of distance to a subway station. The code to do this is shown hereafter. You simply need to replace the original point dataset to the newly generated buffer dataset, as is done using Code Block 6-5.

Code Block 6-5. Create a plot with the buffer data

```python
import contextily as cx

# import the paris subway station data
gpd.io.file.fiona.drvsupport.supported_drivers['KML'] = 'rw'
paris = gpd.read_file('ParisSeineData.kml')
paris = paris.loc[1:,:]

# use paris data to set extent but leave invisible
ax = paris.plot(figsize=(15,15), color="None")

# add the data
data.buffer(0.01).plot(ax=ax, edgecolor='black', color='None')

# add the background map
cx.add_basemap(ax, crs=data.crs)
```

The buffers are shown as black circles on the background map. You can check out the result in Figure 6-9.

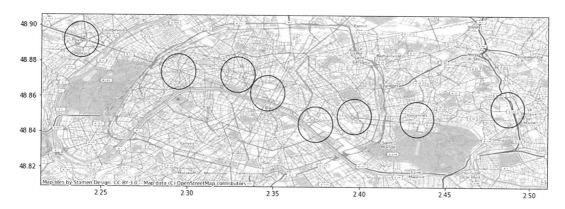

Figure 6-9. *The plot resulting from Code Block 6-5. Image by author using contextily source data and image as referenced in the image*

With this result, we have successfully created a spatial layer to help us in filtering houses to select. Let's now move on to implementing the following two criteria using buffers as well.

Creating Buffers Around Lines in Python

In this second part of the exercise, you will see how to create a buffer around a line. We will take a line feature that shows the railroad of the subway line, and we will create a buffer that we consider too close to the subway line. Imagine that the subway makes quite some noise. For full disclosure, a large part of the line is actually underground, but let's leave that out of the equation here, for the purpose of an easier exercise.

Rather than importing a line dataset, let's use the point dataset from before and convert the eight points into one single line feature. To do so, we first need to make sure that the points are in the right order, and then we pass them to the LineString function from the shapely package. This is all done in the code in Code Block 6-6.

Code Block 6-6. A LineString object

```
from shapely.geometry.linestring import LineString
LineString(data.loc[[7,6,5,4,0,1,2,3], 'geometry'].reset_index(drop=True))
```

When calling this in a notebook, you will see how a line is automatically printed, as shown in Figure 6-10.

Figure 6-10. *The LineString object printed out. Image by author*

This visualization isn't particularly useful, so we'd better try to add this to our existing plot. The code in Code Block 6-7 does exactly that, by storing the LineString as a geopandas dataframe and then plotting it.

Code Block 6-7. Add this LineString to our existing plot

```
import pandas as pd

df = pd.DataFrame(
    {
      'Name': ['metro'],
      'geometry': [LineString(data.loc[[7,6,5,4,0,1,2,3], 'geometry'].reset_
      index(drop=True))]
    }
)

gdf = gpd.GeoDataFrame(df)

gdf
```

You will see that the resulting geodataframe has exactly one line, which is the line representing our subway, as shown in Figure 6-11.

	Name	geometry
0	metro	LINESTRING Z (2.23821 48.89170 0.00000, 2.2949...

Figure 6-11. *The dataframe. Image by author*

To plot the line, let's add this data into the plot with the background map directly, using the code in Code Block 6-8.

Code Block 6-8. Add this data in the plot

```
import contextily as cx

# use paris data to set extent but leave invisible
ax = paris.plot(figsize=(15,15), color="None")

# add the point data
data.buffer(0.01).plot(ax=ax, edgecolor='black', color='None')

# add the line data
gdf.plot(ax = ax, color = 'black')

# add the background map
cx.add_basemap(ax, crs=data.crs)
```

You now obtain a map that has the subway station buffers and the subway rails as a line. The result is shown in Figure 6-12.

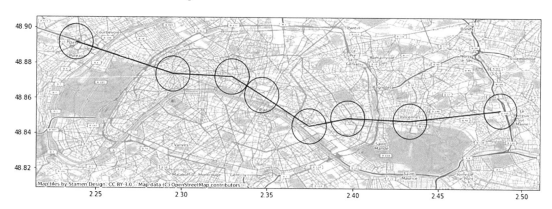

Figure 6-12. *The plot with the two data types. Image by author using contextily source data and image as referenced in the image*

The next step is to compute a buffer around this line to indicate an area that you want to deselect for your search for a house, to respect the criteria given in the introduction. This can be done using the same operation as used before, but now we will choose a slightly smaller buffer size to avoid deselecting too much areas. This is done in Code Block 6-9.

Code Block 6-9. Adding a smaller buffer

```
gdf.buffer(0.001)
```

By creating the buffer in this way, you end up with a geodataframe that contains a polygon of the buffer, rather than with the initial line data. This is shown in Figure 6-13.

```
0      POLYGON ((2.29510 48.87480, 2.32941 48.87326, ...
dtype: geometry
```

Figure 6-13. *The polygon with the buffer. Image by author*

Now, you can simply add this polygon in the plot, and you'll obtain a polygon that shows areas that you should try to find a house in and some subareas that should be avoided. Setting the transparency using the alpha parameter can help a lot to make more readable maps. This is done in Code Block 6-10.

Code Block 6-10. Improve the plot

```
import contextily as cx

# use paris data to set extent but leave invisible
ax = paris.plot(figsize=(15,15), color="None")

# add the point data
data.buffer(0.01).plot(ax=ax, edgecolor='black', color='green', alpha=0.5)

# add the line data
gdf.buffer(0.001).plot(ax = ax, edgecolor='black', color = 'red',
alpha = 0.5)

# add the background map
cx.add_basemap(ax, crs=data.crs)
```

You will obtain the intermediate result shown in Figure 6-14.

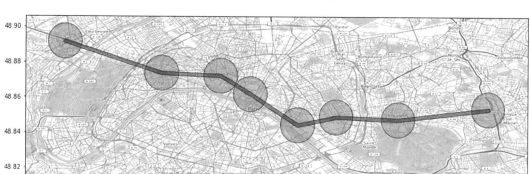

Figure 6-14. *The plot resulting from Code Block 6-10. Image by author using*
contextily source data and image as referenced in the image

This shows the map of Paris in which the best circles for use are marked in green, but
in which the red polygon should be avoided as it is too close to the subway line. In the
following section, we will add a third criterion on the map: proximity to a park. This will
be done by creating buffers on polygons.

Creating Buffers Around Polygons in Python

The third criterion for the house selection is proximity to a park. In the Paris_Parks.kml
dataset, you can find some parks in Paris. This data is just to serve the example, it is far
from perfect, but it will do the trick for this exercise. You can import the data using the
code in Code Block 6-11.

Code Block 6-11. Importing the parks data

```
# import the paris parks data
gpd.io.file.fiona.drvsupport.supported_drivers['KML'] = 'rw'
parks = gpd.read_file('Paris_Parks.kml')
parks
```

In Figure 6-15, you will see that there are 18 parks in this dataset, all identified as polygons.

	Name	Description	geometry
0	Park 1		POLYGON Z ((2.23251 48.86778 0.00000, 2.22882 ...
1	Park 2		POLYGON Z ((2.28578 48.86047 0.00000, 2.28754 ...
2	Park 3		POLYGON Z ((2.28848 48.85499 0.00000, 2.29230 ...
3	Park 4		POLYGON Z ((2.31160 48.86270 0.00000, 2.31118 ...
4	Park 5		POLYGON Z ((2.31043 48.86911 0.00000, 2.31015 ...
5	Park 6		POLYGON Z ((2.32140 48.86372 0.00000, 2.32994 ...
6	Park 7		POLYGON Z ((2.33771 48.86622 0.00000, 2.33650 ...
7	Park 8		POLYGON Z ((2.30552 48.87962 0.00000, 2.30653 ...
8	Park 9		POLYGON Z ((2.35484 48.84382 0.00000, 2.35595 ...
9	Park 10		POLYGON Z ((2.37595 48.83865 0.00000, 2.38428 ...
10	Park 11		POLYGON Z ((2.38680 48.84394 0.00000, 2.38797 ...
11	Park 12		POLYGON Z ((2.38598 48.84260 0.00000, 2.38652 ...
12	Park 13		POLYGON Z ((2.39631 48.84879 0.00000, 2.39573 ...
13	Park 14		POLYGON Z ((2.39779 48.83172 0.00000, 2.39642 ...
14	Park 15		POLYGON Z ((2.46557 48.85144 0.00000, 2.47111 ...
15	Park 16		POLYGON Z ((2.46759 48.86429 0.00000, 2.47261 ...
16	Park 17		POLYGON Z ((2.47823 48.84859 0.00000, 2.47857 ...
17	Park 18		POLYGON Z ((2.45078 48.85457 0.00000, 2.45756 ...

Figure 6-15. *The data from Code Block 6-11. Image by author*

You can visualize this data directly inside our map, by adding it as done in Code Block 6-12.

Code Block 6-12. Visualize the data directly in our map

```
import contextily as cx

# use paris data to set extent but leave invisible
ax = paris.plot(figsize=(15,15), color="None")

# add the point data
data.buffer(0.01).plot(ax=ax, edgecolor='black', color='green', alpha=0.5)

# add the line data
gdf.buffer(0.001).plot(ax = ax, edgecolor='black', color = 'red',
alpha = 0.5)

# add the parks
parks.plot(ax=ax, edgecolor='black', color="none")

# add the background map
cx.add_basemap(ax, crs=data.crs)
```

The parks are shown in the map as black contour lines. No buffers have yet been created. This intermediate result looks as shown in Figure 6-16.

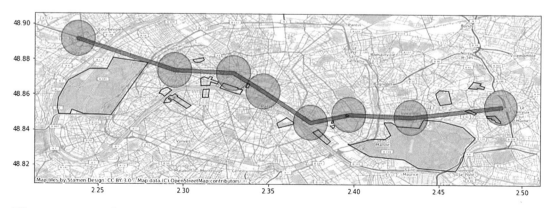

Figure 6-16. *The map with the parks added to it. Image by author using contextily source data and image as references in the image*

Of course, it is unlikely that you will find a house inside a park, so we need to make our search area such that it takes into account a border around those parks. This, again, can be done by adding a buffer to our polygon data. The buffer operation works just like it did before, by calling buffer with a distance. This is done in Code Block 6-13.

Code Block 6-13. Adding the buffer to the parks

```
parks.buffer(0.01)
```

This looks like shown in Figure 6-17.

```
0     POLYGON ((2.22785 48.87663, 2.24030 48.88317, ...
1     POLYGON ((2.27757 48.86618, 2.27899 48.86822, ...
2     POLYGON ((2.27980 48.85996, 2.28117 48.86236, ...
3     POLYGON ((2.31092 48.87267, 2.31461 48.87293, ...
4     POLYGON ((2.30602 48.87809, 2.30767 48.87890, ...
5     POLYGON ((2.31340 48.86973, 2.31544 48.87244, ...
6     POLYGON ((2.33971 48.87602, 2.34109 48.87574, ...
7     POLYGON ((2.30353 48.88942, 2.30973 48.89068, ...
8     POLYGON ((2.34991 48.85252, 2.35571 48.85580, ...
9     POLYGON ((2.37151 48.84761, 2.37464 48.84917, ...
10    POLYGON ((2.37680 48.84407, 2.37681 48.84490, ...
11    POLYGON ((2.38020 48.83420, 2.38001 48.83434, ...
12    POLYGON ((2.40277 48.85642, 2.40307 48.85617, ...
13    POLYGON ((2.39755 48.84199, 2.40753 48.84457, ...
14    POLYGON ((2.46230 48.86089, 2.46474 48.86174, ...
15    POLYGON ((2.45988 48.87066, 2.46284 48.87424, ...
16    POLYGON ((2.47697 48.83768, 2.47672 48.83772, ...
17    POLYGON ((2.44181 48.85015, 2.43975 48.85433, ...
```

Figure 6-17. *The data resulting from Code Block 6-13*

After the buffer, you have polygon data, just like you had before. Yet the size of the polygon is now larger as it also has the buffers around the original polygons. Let's now add this into our plot, to see how this affects the places in which we want to find a house. This is done in Code Block 6-14.

Code Block 6-14. Adding all the data together

```
import contextily as cx

# use paris data to set extent but leave invisible
ax = paris.plot(figsize=(15,15), color="None")

# add the point data
data.buffer(0.01).plot(ax=ax, edgecolor='none', color='yellow', alpha=0.5,
zorder=2)

# add the line data
gdf.buffer(0.001).plot(ax = ax, edgecolor='black', color = 'red', alpha =
1, zorder=3)

# add the parks
parks.buffer(0.01).plot(ax=ax, edgecolor='none', color="green",
alpha = 0.5)

# add the background map
cx.add_basemap(ax, crs=data.crs)
```

The colors and "zorder" (order of overlay) have been adjusted a bit to make the map more readable. After all, it starts to contain a large number of features. You will see the result shown in Figure 6-18.

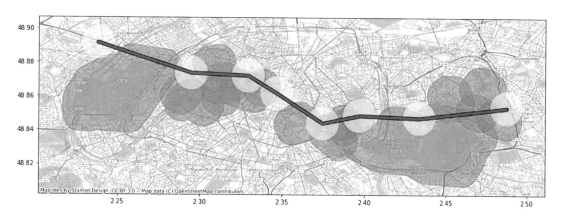

Figure 6-18. *The plot resulting from Code Block 6-14. Image by author using contextily source data and image as referenced in the image*

This map is a first result that you could use. Of course, you could go even further and combine this with the theory from Chapter 5, in which you have learned how to use operations from set theory to combine different shapes. Let's see how to do this, with a final goal to obtain a dataframe that only contains the areas in which we do want to find a house, based on all three criteria from the introduction.

Combining Buffers and Set Operations

To combine the three buffers together, we will need multiple operations. We first need to do an intersection between "subway station proximity" and "park proximity." After all, we want to have both of them and cannot suffice having one or the other. The intersection is the right operation for this, as you have seen in the previous chapter.

The code in Code Block 6-15 creates an intersection layer between the two buffered dataframes. They are first assigned to an individual dataframe variable each, to create easier-to-use objects.

Code Block 6-15. Prepare to create an intersection layer

```
station_buffer = data.buffer(0.01)
rails_buffer = gdf.buffer(0.001)
park_buffer = parks.buffer(0.01)

A = gpd.GeoDataFrame({'geometry': station_buffer})
B = gpd.GeoDataFrame({'geometry': park_buffer})
C = gpd.GeoDataFrame({'geometry': rails_buffer})
```

Then an overlay method is applied, using the intersection parameter, as shown in Code Block 6-16.

Code Block 6-16. Create the intersection layer

```
A_and_B = A.overlay(B, how='intersection')
A_and_B
```

You will obtain a dataset that looks like the data shown in Figure 6-19.

	geometry
0	POLYGON ((2.36976 48.83550, 2.36892 48.83600, …
1	POLYGON ((2.35308 48.85417, 2.35229 48.85359, …
2	POLYGON ((2.38442 48.84334, 2.38428 48.84236, …
3	POLYGON ((2.39404 48.83929, 2.39312 48.83962, …
4	POLYGON ((2.38442 48.84334, 2.38428 48.84236, …
5	POLYGON ((2.39597 48.83891, 2.39500 48.83905, …
6	POLYGON ((2.38442 48.84334, 2.38428 48.84236, …
7	POLYGON ((2.39793 48.83891, 2.39695 48.83886, …
8	POLYGON ((2.40526 48.84330, 2.40468 48.84252, …
9	POLYGON ((2.40526 48.84330, 2.40468 48.84252, …
10	POLYGON ((2.44327 48.84639, 2.44313 48.84542, …

Figure 6-19. *The data resulting from Code Block 6-16. Image by author*

With this intersection of stations and parks, we now need to remove all locations that are too close to a subway line, as this is expected to be too noisy, as specified in the introduction.

To do this, we can also use an overlay, but this time we do not need the intersection from set theory because an intersection would leave us with all places that have station, park, and railway proximity. However, what we want is station and park proximity, but not railway proximity. For this, we need to use the difference operation from set theory. The code in Code Block 6-17 shows how this can be done.

Code Block 6-17. Creating the difference

```
A_and_B_not_C = A_and_B.overlay(C, how='difference')
A_and_B_not_C
```

The data still looks like a dataframe from before. The only difference that occurs is that the data becomes much more complex with every step, as the shapes of our acceptable locations become less and less regular. Let's do a map of our final object using Code Block 6-18.

Code Block 6-18. Create a map of the final object

```
import contextily as cx

# use paris data to set extent but leave invisible
ax = paris.plot(figsize=(15,15), edgecolor="none", color="none")

A_and_B_not_C.plot(ax=ax, edgecolor='none', color='green', alpha=0.8)

# add the background map
cx.add_basemap(ax, crs=data.crs)
```

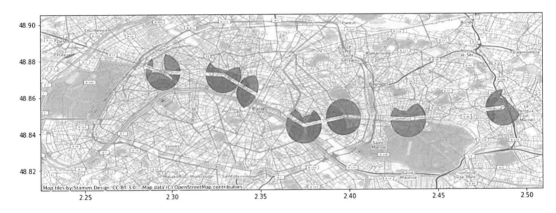

Figure 6-20. *The final map of the exercise. Image by author using contextily source data and image as referenced in the image*

As you can see in Figure 6-20, the green areas are now a filter that we could use to select houses based on coordinates. This answers the question posed in the exercise and results in an interesting map as well. If you want to go further with this exercise, you could create a small dataset containing point data for houses. Then, for looking up whether a house (point data coordinate) is inside a polygon, you can use the operation that is called "contains" or "within." Documentation can be found here:

- https://geopandas.org/en/stable/docs/reference/api/
  geopandas.GeoSeries.within.html

- https://geopandas.org/en/stable/docs/reference/api/
  geopandas.GeoSeries.contains.html

This operation is left as an exercise, as it goes beyond the demonstration of the buffer operation, which is the focus of this chapter.

Key Takeaways

1. There are numerous basic geodata operations that are standardly implemented in most geodata tools. They may seem simple at first sight, but applying them to geodata can come with some difficulties.

2. The buffer operation adds a polygon around a vector object. Whether the initial object is point, line, or polygon, the result is always a polygon.

3. When applying a buffer, one can choose the distance of the buffer's boundary to the initial object. The choice depends purely on the use case.

4. Once buffers are computed, they can be used for mapping purposes, or they can be used in further geospatial operations.

CHAPTER 7

Merge and Dissolve

In the previous two chapters, we have started looking at a number of common geospatial operations: data operations that are not possible, or at least not common, on regular data, but that are very common on geospatial data.

Although this book is working toward machine learning applications, it is good to have an understanding of the types of geospatial operations that exist and have a working understanding that will allow you to figure out other, comparable operations using documentation or research. The standard operations that are showcased in this book are

- Clipping and intersecting

- Buffering

- Merge and dissolve

- Erase

In the previous chapter, you have already seen clipping, intersecting, and buffering. You have even seen a use case that combines those methods in a logical manner.

In this chapter, we will look at merging and dissolving, which are again very different operations than those previously presented. They are also among the standard geospatial operations, and it is useful to master these tools.

As explained in previous chapters, those spatial operations should all be seen as tools in a toolkit, and when you want to achieve a specific goal, you need to select the different tools that you need to get there. This often means using different combinations of tools in an intelligent manner. The more tools you know, the more you will be able to achieve.

We will start this chapter with the merge operation, covering both theory and implementation in Python, and we will then move on to the dissolve operation. At the end, you will see a comparison of both.

© Joos Korstanje 2022
J. Korstanje, *Machine Learning on Geographical Data Using Python*,
https://doi.org/10.1007/978-1-4842-8287-8_7

The Merge Operation

Let's begin by introducing the merge operation. We will start with a theoretical introduction and some examples and then move on to the Python implementation of the merge operation.

What Is a Merge?

Merging geodata, just like with regular data, consists of taking multiple input datasets and making them into a single new output feature. In the previous chapter, you already saw a possible use case for a merge. If you remember, multiple "suitability" polygons were created based on multiple criteria. At the end, all of these polygons were combined into a single spatial layer. Although another solution was used in that example, a merge could have been used to get all those layers together in one and the same layer.

A Schematic Example of Merging

Let's clarify this definition using a visual example. As a simple and intuitive example, imagine that you have three datasets containing some polygon features, each on a specific region. You are asked to make a map of the three regions combined, and you therefore need to find a way to put those three datasets together into a single, three-region, dataset. The merge operation will do this for you. Figure 7-1 shows you how this works.

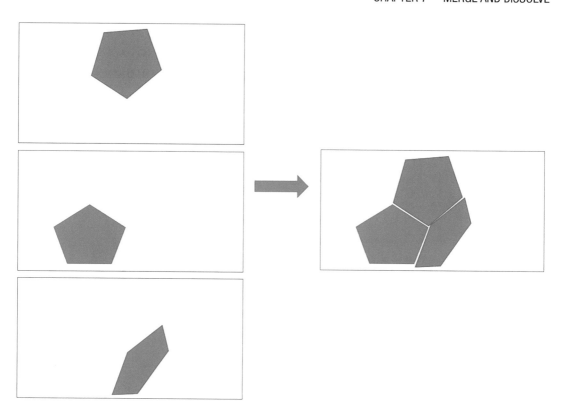

Figure 7-1. *The merge operation. Image by author*

Different Definitions of Merging

The definition of the merge operation is not exactly the same depending on the software that you use. It is important to understand how different types of merge work, so that you can find the one you need based on their descriptions in the documentation.

The real definition of the merge operation is the one in which you append features together, row-wise. This does not change anything to the existing features and does not affect anything column-wise. This is what is shown in the schematic drawing in the example as well.

The word merge can sometimes be used for indicating joins. It would have been better to call these operations "join," but some software have not made this choice. There are two types of joins that are common: spatial joins and attribute joins.

Attribute joins are probably the type of join that you are already aware of. This is an SQL-like join in which your geodata has some sort of key that is used to add information from a second table that has this key as well. An example is shown in the schematic drawing in Figure 7-2.

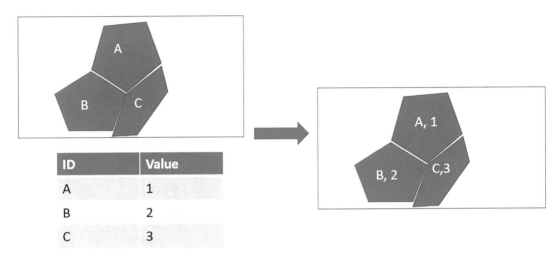

Figure 7-2. *Attribute join. Image by author*

As you can see, this is a simple SQL-like join that uses a common identifier between the two datasets to add the columns of the attribute table into the columns of the geodata dataset.

An alternative is the spatial join, which is a bit more complex. The spatial join also combines columns of two datasets, but rather than using a common identifier, it uses the geographic coordinates of the two datasets. The schematic drawing in Figure 7-3 shows how this can be imagined.

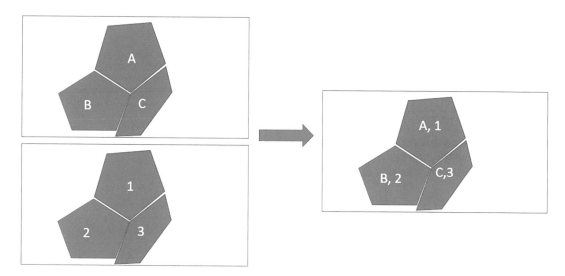

Figure 7-3. *Spatial join. Image by author*

In this example, the spatial join is relatively easy, as the objects are exactly the same in both input datasets. In reality, you may well see slight differences in the features, but you may also have different features that you want to join. You can specify all types of spatial join parameters to make the right combination:

- Joining all objects that are near each other (specify a distance)

- Joining based on one object containing the other

- Joining based on intersections existing

This gives you a lot of tools to work with for combining datasets together, both row-wise (merge) and column-wise (join). Let's now see some examples of the merge, attribute join, and spatial join in Python.

Merging in Python

In the coming examples, we will be looking at some easy-to-understand data. There are multiple small datasets, and throughout the exercise, we will do all the three types of merges.

The data contains

- A file with three polygons for Canada, USA, and Mexico

- A file with some cities of Canada

- A file with some cities of the USA

- A file with some cities of Mexico

During the exercise, we will take the following steps:

- Combine the three city files using a row-wise merge

- Add a new variable to the combined city file using an attribute lookup

- Find the country of each of the cities using a spatial lookup with the polygon file

Let's now start by combining the three city files into a single layer with all the cities combined.

Row-Wise Merging in Python

The row-wise merge operation is the operation that is generally referred to when talking about the merge operation. It is like a row-wise concatenation of the three datasets. Let's see how to do this in Python using the three example data files containing some of the cities of each country. Let's start by importing the data as follows and plot each of them separately and add a background map, using the functionalities that you have already seen in earlier chapters. This is shown in Code Block 7-1.

Code Block 7-1. Importing the data

```
import geopandas as gpd
import fiona

gpd.io.file.fiona.drvsupport.supported_drivers['KML'] = 'rw'
us_cities = gpd.read_file('/kaggle/input/chapter7/USCities.kml')
us_cities
```

The US cities look as shown in Figure 7-4.

	Name	Description	geometry
0	Las Vegas		POINT Z (-117.05255 34.66010 0.00000)
1	New York		POINT Z (-75.21057 40.87178 0.00000)
2	Washington		POINT Z (-78.37464 38.81382 0.00000)

Figure 7-4. *The US cities*

Let's import the Canada cities as shown in Code Block 7-2.

Code Block 7-2. Importing the Canada cities

```
canada_cities = gpd.read_file('/kaggle/input/chapter7/CanadaCities.kml')
canada_cities
```

They look as shown in Figure 7-5.

	Name	Description	geometry
0	Toronto		POINT Z (-80.27265 44.26100 0.00000)
1	Quebec		POINT Z (-71.52753 47.08515 0.00000)
2	Montreal		POINT Z (-74.64765 46.02755 0.00000)
3	Vancouver		POINT Z (-123.30126 49.72246 0.00000)

Figure 7-5. *The Canada cities. Image by author*

The Mexico cities are imported using Code Block 7-3.

Code Block 7-3. Importing the Mexico cities

```
mexico_cities = gpd.read_file('/kaggle/input/chapter7/MexicoCities.kml')
mexico_cities
```

They look as shown in Figure 7-6.

	Name	Description	geometry
0	Guadalajara		POINT Z (-103.10867 20.79000 0.00000)
1	Mexico City		POINT Z (-99.37332 19.75946 0.00000)

Figure 7-6. *The Mexico cities. Image by author*

We can create a map of all three of those datasets using the syntax that you have seen earlier in this book. This is done in Code Block 7-4.

Code Block 7-4. Creating a map of the datasets

```
import contextily as cx

# us cities
ax = us_cities.plot(markersize=128, figsize=(15,15))

# canada cities
canada_cities.plot(ax=ax, markersize=128)

# mexico cities
mexico_cities.plot(ax = ax, markersize=128)

# contextily basemap
cx.add_basemap(ax, crs=us_cities.crs)
```

This will result in the map shown in Figure 7-7.

Figure 7-7. *The map created using Code Block 7-4. Image by author using contextily source data and image as referenced in the image*

Now, this is not too bad already, but we actually want to have all this data in just one layer, so that it is easier to work with. To do so, we are going to do a row-wise merge operation. This can be done in Python using the pandas concat method. It is shown in Code Block 7-5.

Code Block 7-5. Using concatenation

```
import pandas as pd
cities = pd.concat([us_cities, canada_cities, mexico_cities])
cities
```

You will obtain a dataset, in which all the points are now combined. Cities now contain the rows of all the cities of the three input geodataframes, as can be seen in Figure 7-8.

	Name	Description	geometry
0	Las Vegas		POINT Z (-117.05255 34.66010 0.00000)
1	New York		POINT Z (-75.21057 40.87178 0.00000)
2	Washington		POINT Z (-78.37464 38.81382 0.00000)
0	Toronto		POINT Z (-80.27265 44.26100 0.00000)
1	Quebec		POINT Z (-71.52753 47.08515 0.00000)
2	Montreal		POINT Z (-74.64765 46.02755 0.00000)
3	Vancouver		POINT Z (-123.30126 49.72246 0.00000)
0	Guadalajara		POINT Z (-103.10867 20.79000 0.00000)
1	Mexico City		POINT Z (-99.37332 19.75946 0.00000)

Figure 7-8. *The concatenated dataframe. Image by author*

If we now plot this data, we just have to plot one layer, rather than having to plot three times. This is done in Code Block 7-6. You can see that it has all been successfully merged into a single layer.

Code Block 7-6. Plotting the concatenated cities

```
ax = cities.plot(markersize=128,figsize=(15,15))
cx.add_basemap(ax, crs=us_cities.crs)
```

This looks as is shown in Figure 7-9.

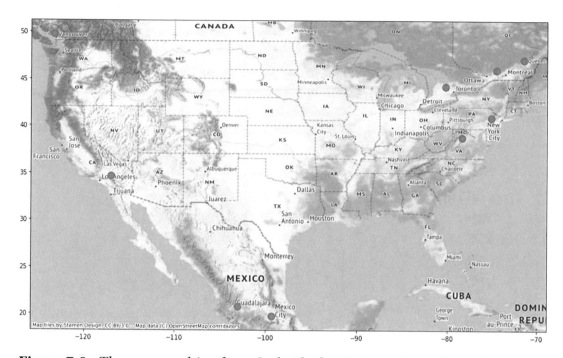

Figure 7-9. *The map resulting from Code Block 7-6. Image by author using contextily source data and image as referenced in the image*

You can also see that all points now have the same color, because they are now all on one single dataset. This fairly simple operation of row-wise merging will prove to be very useful in your daily GIS operations.

Now that we have combined all data into one layer, let's add some features using an attribute join.

Attribute Join in Python

To add a new variable based on a lookup table and an SQL-like join, we are going to use an attribute join. To do this, we first generate a lookup table with some fictitious data, just to see how to make the operations work. This is done in Code Block 7-7.

Code Block 7-7. Create a lookup table

```
lookup = pd.DataFrame({
    'city': [
        'Las Vegas',
        'New York',
```

```
        'Washington',
        'Toronto',
        'Quebec',
        'Montreal',
        'Vancouver',
        'Guadalajara',
        'Mexico City'
    ],
    'population': [
        1234,
        2345,
        3456,
        4567,
        4321,
        5432,
        6543,
        1357,
        2468

    ]
})

lookup
```

The lookup table looks as shown in Figure 7-10.

	city	population
0	Las Vegas	1234
1	New York	2345
2	Washington	3456
3	Toronto	4567
4	Quebec	4321
5	Montreal	5432
6	Vancouver	6543
7	Guadalajara	1357
8	Mexico City	2468

Figure 7-10. *The lookup table. Image by author*

Now, to add this data into the geodataframe, we want to do an SQL-like join, in which the column "population" from the lookup table is added onto the geodataframe based on the column "Name" in the geodataframe and the column "city" in the lookup table. This can be accomplished using Code Block 7-8.

Code Block 7-8. Attribute join

```
cities_new = cities.merge(lookup, left_on='Name', right_on='city')
cities_new
```

You can see in the dataframe that the population column has been added, as is shown in Figure 7-11.

	Name	Description	geometry	city	population
0	Las Vegas	POINT Z (-117.05255 34.66010 0.00000)	Las Vegas	1234
1	New York	POINT Z (-75.21057 40.87178 0.00000)	New York	2345
2	Washington	POINT Z (-78.37464 38.81382 0.00000)	Washington	3456
3	Toronto	POINT Z (-80.27265 44.26100 0.00000)	Toronto	4567
4	Quebec	POINT Z (-71.52753 47.08515 0.00000)	Quebec	4321
5	Montreal	POINT Z (-74.64765 46.02755 0.00000)	Montreal	5432
6	Vancouver	POINT Z (-123.30126 49.72246 0.00000)	Vancouver	6543
7	Guadalajara	POINT Z (-103.10867 20.79000 0.00000)	Guadalajara	1357
8	Mexico City	POINT Z (-99.37332 19.75946 0.00000)	Mexico City	2468

Figure 7-11. *The data resulting from Code Block 7-8. Image by author*

You can now access this data easily, for example, if you want to do filters, computations, etc. Another example is to use this attribute data to adjust the size of each point on the map, depending on the (simulated) population size (of course, this is toy data so the result is not correct, but feel free to improve on this if you want to). The code is shown in Code Block 7-9.

Code Block 7-9. Plot the new data

```
ax = cities_new.plot(markersize=cities_new['population'] // 10,
figsize=(15,15))
cx.add_basemap(ax, crs=us_cities.crs)
```

The result in Figure 7-12 shows the cities' sizes being adapted to the value in the column population, which was added to the dataset through an attribute join.

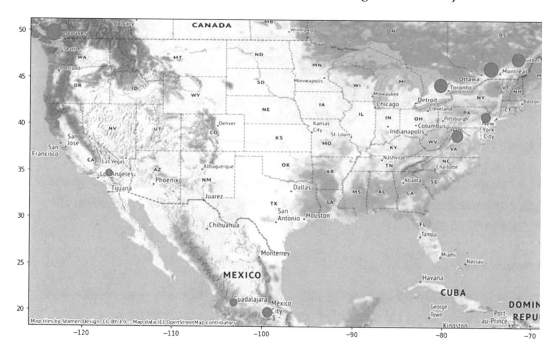

Figure 7-12. *The map resulting from Code Block 7-9. Image by author using contextily source data and image as referenced in the image*

Spatial Join in Python

You have now seen how to join two datasets based on tabular attributes. Yet it is also possible to join data on spatial information. For this example, we will use the dataset containing approximated country polygons, which can be found in the KML file. Again, the goal is not to provide final products, but rather to show the techniques to make them, so the data is a simulated approximation of what the real data would look like. You can import the country polygons using the code in Code Block 7-10.

Code Block 7-10. Importing the data

```
countries = gpd.read_file('NorthMiddleAmerciaCountries.kml')
countries
```

The data is shown in Figure 7-13.

	Name	Description	geometry
0	US	POLYGON Z ((-124.08614 48.59385 0.00000, -124....
1	Canada	POLYGON Z ((-140.72136 69.80715 0.00000, -141....
2	Mexico	POLYGON Z ((-117.20749 32.40949 0.00000, -105....

Figure 7-13. *The data resulting from Code Block 7-10*

If we plot the data against the background map, you can see that the polygons are quick approximations of the countries' borders, just for the purpose of this exercise. This is done in Code Block 7-11.

Code Block 7-11. Plotting the data

```
ax = countries.plot(figsize=(15,15), edgecolor='black', facecolor='none')
cx.add_basemap(ax, crs=countries.crs)
```

This will show the map in Figure 7-14.

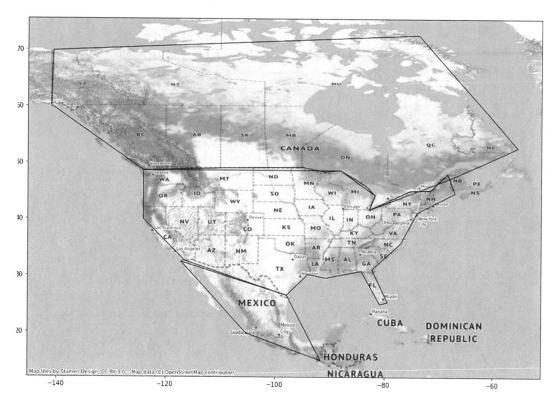
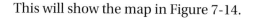

Figure 7-14. *The plot resulting from Code Block 7-11. Image by author using contextily source data and image as referenced in the image*

You can see some distortion on this map. If you have followed along with the theory on coordinate systems in Chapter 2, you should be able to understand where that is coming from and have the tools to rework this map's coordinate system if you'd want to. For the current exercise, those distortions are not a problem. Now, let's add our cities onto this map, using Code Block 7-12.

Code Block 7-12. Add the cities to the map

```
ax = countries.plot(figsize=(15,15), edgecolor='black', facecolor='none')
cities_new.plot(ax=ax, markersize=cities_new['population'] // 10,
figsize=(15,15))
cx.add_basemap(ax, crs=countries.crs)
```

This gives the combined map shown in Figure 7-15.

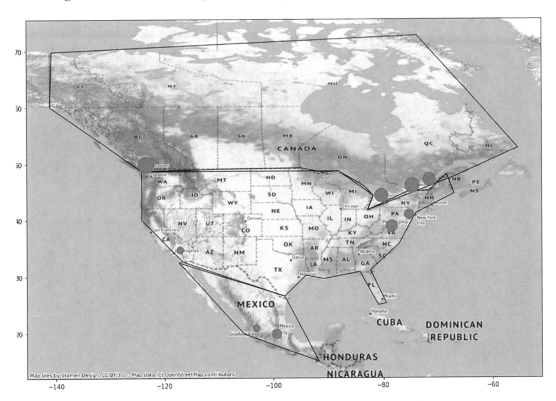

Figure 7-15. *The combined map. Image by author using contextily source data and image as referenced in the image*

This brings us to the topic of the spatial join. In this map, you see that there are two datasets:

- The cities only contain information about the name of the city and the population.

- The countries are just polygons.

It would be impossible to use an SQL-like join to add a column country to each of the rows in the city dataset. However, we can clearly see that based on the spatial information, it is possible to find out in which country each of the cities is located.

The spatial join is made exactly for this purpose. It allows us to combine two datasets column-wise, even when there is no common identifier: just based on spatial information. This is one of those things that can be done with geodata but not with regular data.

You can see in Code Block 7-13 how a spatial join is done between the cities and countries datasets, based on a "within" spatial join: the city needs to be inside the polygon to receive its attributes.

Code Block 7-13. Spatial join between the cities and the countries

```
cities_3 = cities_new.sjoin(countries, how="inner", predicate='within')
cities_3
```

The data looks as shown in Figure 7-16.

Name_left	Description_left	geometry	city	population	index_right	Name_right	Description_right
Las Vegas		POINT Z (-117.05255 34.66010 0.00000)	Las Vegas	1234	0	US	
New York		POINT Z (-75.21057 40.87178 0.00000)	New York	2345	0	US	
Washington		POINT Z (-78.37464 38.81382 0.00000)	Washington	3456	0	US	
Toronto		POINT Z (-80.27265 44.26100 0.00000)	Toronto	4567	1	Canada	
Quebec		POINT Z (-71.52753 47.08515 0.00000)	Quebec	4321	1	Canada	
Montreal		POINT Z (-74.64765 46.02755 0.00000)	Montreal	5432	1	Canada	
Vancouver		POINT Z (-123.30126 49.72246 0.00000)	Vancouver	6543	1	Canada	
Guadalajara		POINT Z (-103.10867 20.79000 0.00000)	Guadalajara	1357	2	Mexico	
Mexico City		POINT Z (-99.37332 19.75946 0.00000)	Mexico City	2468	2	Mexico	

Figure 7-16. *The data resulting from Code Block 7-13. Image by author*

You see that the name of the country has been added to the dataset of the cities. We can now use this attribute for whatever we want to in the cities dataset. As an example, we could give the points a color based on their country, using Code Block 7-14.

Code Block 7-14. Colors based on country

```
cities_3['color'] = cities_3['index_right'].map({0: 'green', 1: 'yellow', 2: 'blue'})

ax = cities_3.plot(markersize=cities_3['population'] // 10, c=cities_3['color'], figsize=(15,15))
cx.add_basemap(ax, crs=cities_3.crs)
```

This results in the map shown in Figure 7-17.

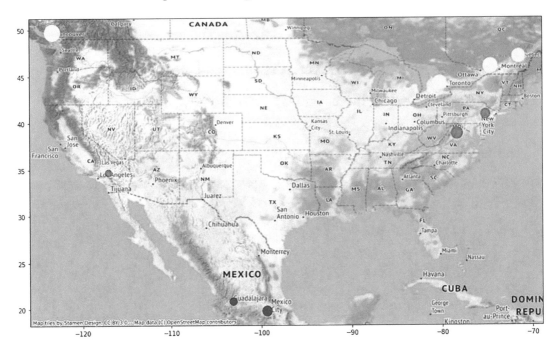

Figure 7-17. *The map resulting from Code Block 7-14. Image by author using contextily source data and image as referenced in the image*

With this final result, you have now seen multiple ways to combine datasets into a single dataset:

- The row-wise concatenation operation generally called merge in GIS

- The attribute join, which is done with a geopandas method confusingly called merge, whereas it is generally referred to as a join rather than a merge

- The spatial join, which is a join that bases itself on spatial attributes rather than on any common identifier

In the last part of this chapter, you'll discover the dissolve operation, which is often useful in case of joining many datasets.

The Dissolve Operation

When combining many different datasets, it can often happen that you obtain a lot of overlapping features, like polygons of different granularity being all kept in the data, or the same features being present in each of the dataset and creating doubled information. The dissolve operation is something that can solve such problems.

What Is the Dissolve Operation?

The dissolve operation is a tool inside a larger family of generalization tools. They allow you to combine data that is too detailed or too granular in larger features. You can see the dissolve tool like a grouping operation. It works in a similar fashion as the groupby operation in SQL or pandas.

Schematic Overview of the Dissolve Operation

You can see a schematic drawing of a dissolve operation in Figure 7-18. In this image, you see that multiple polygons are being grouped into one, based on their numerical feature, as shown in Figure 7-18.

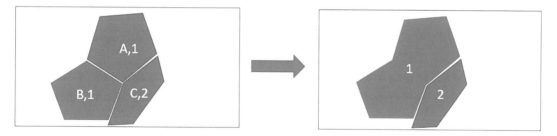

Figure 7-18.   *The dissolve operation. Image by author*

The polygons A and B both have the value 1, so grouping by value would combine those two polygons into one polygon. This operation can be useful when your data is too granular, which may be because you have done a lot of geospatial operations or may be because you have merged a large number of data files.

The Dissolve Operation in Python

Let's now see how to do a dissolve operation in Python. Let's use the same data as in the previous example and try to see how we can use the dissolve operation to regroup the polygons based on them being in North America (USA and Canada) or in Middle America (Mexico). To do so, we have to add an attribute on the countries dataset that indicates this. The code in Code Block 7-15 shows how this can be done.

Code Block 7-15. Add a country attribute

```
countries['Area'] = ['North America', 'North America', 'Middle America']
countries
```

Once you execute this, you'll see that a new column has been added to the dataset, as shown in Figure 7-19.

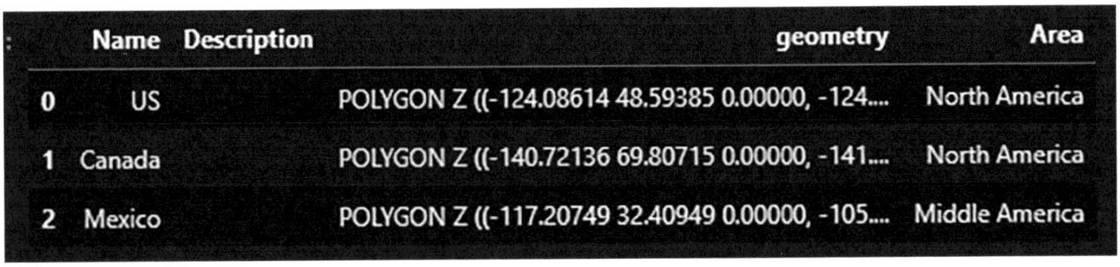

	Name	Description	geometry	Area
0	US	POLYGON Z ((-124.08614 48.59385 0.00000, -124....	North America
1	Canada	POLYGON Z ((-140.72136 69.80715 0.00000, -141....	North America
2	Mexico	POLYGON Z ((-117.20749 32.40949 0.00000, -105....	Middle America

Figure 7-19. *The data resulting from Code Block 7-15. Image by author*

Now the goal is to create two polygons: one for North America and one for Middle America. We are going to use the dissolve method for this, as shown in Code Block 7-16.

Code Block 7-16. Dissolve operation

```
areas = countries.dissolve(by='Area')[['geometry']]
areas
```

The newly created dataset is the grouped (dissolved) result of the previous dataset, as shown in Figure 7-20.

	geometry
Area	
Middle America	POLYGON Z ((-117.20749 32.40949 0.00000, -105....
North America	POLYGON Z ((-66.95723 44.61098 0.00000, -73.98...

Figure 7-20. *The grouped dataset. Image by author*

We can now plot this data to see what it looks like, using Code Block 7-17.

Code Block 7-17. Plot the data

```
ax = areas.plot(figsize=(15,15), edgecolor='black', facecolor='none')
cx.add_basemap(ax, crs=areas.crs)
```

The map in Figure 7-21 shows the result of the dissolve operation.

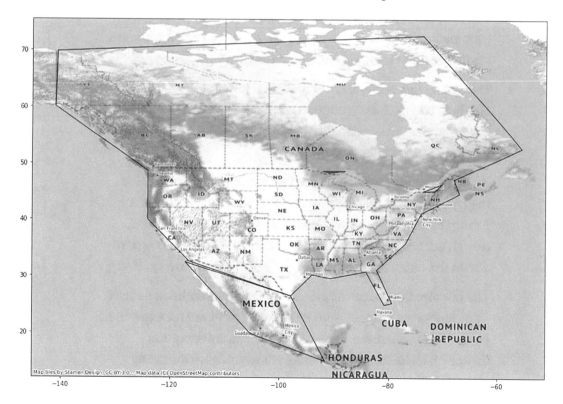

Figure 7-21. *The result of the dissolve operation. Image by author using contextily source data and image as referenced in the image*

This combined result has been grouped by the feature area, and it is a generalized version of the input data. The dissolve operation is therefore much like a groupby operation, which is a very useful tool to master when working with geodata.

Key Takeaways

1. There are numerous basic geodata operations that are standardly implemented in most geodata tools. They may seem simple at first sight, but applying them to geodata can come with some difficulties.

2. The merge operation is generally used to describe a row-wise merge, in which multiple objects are concatenated into one dataset.

3. The attribute join, which is confusingly called merge in geopandas, does a column-wise, SQL-like join using a common attribute between the two input datasets.

4. The spatial join is another column-wise join that allows to combine two datasets without having any common identifier. The correspondence between the rows of the two datasets is done purely by spatial information.

5. Within the spatial join, there are many possibilities to identify the type of spatial relationship that you want to use for joining. Examples are presence of intersections between two objects, one object being totally within the other, one object being partly within the other, or even just two objects being relatively close.

6. The dissolve tool is a tool that allows you to generalize a dataset by grouping your objects into larger objects, based on a specified column. This operation is much like a groupby operation for spatial features, and it is useful when you have too many objects, or too granular objects, that you need to generalize into larger chunks.

CHAPTER 8

Erase

In this chapter, you will learn about the erase operation. The previous three chapters have presented a number of standard GIS operations. Clipping and intersecting were covered in Chapter 5, buffering in Chapter 6, and merge and dissolve were covered in Chapter 7.

This chapter will be the last of those four chapters covering common tools for geospatial analysis. Even though there are much more tools available in the standard GIS toolbox, the goal here is to give you a good mastering of the basics and allowing you to be autonomous in learning the other GIS operations in Python.

The chapter will start with a theoretical introduction of the erase operation and then follow through with a number of example implementations in Python for applying the erase on different geodata types.

The Erase Operation

Erasing is not just an operation in GIS analysis but also a common synonym for deleting. To clarify, let's first look at a "standard" deletion of a feature, which is not a spatial erase, but just a deletion of a complete feature based on its ID rather than based on its spatial information.

Whether you are working with points, lines, or polygons, you could imagine that you simply want to drop one of them. The schematic drawing in Figure 8-1 shows what this would mean.

© Joos Korstanje 2022
J. Korstanje, *Machine Learning on Geographical Data Using Python*,
https://doi.org/10.1007/978-1-4842-8287-8_8

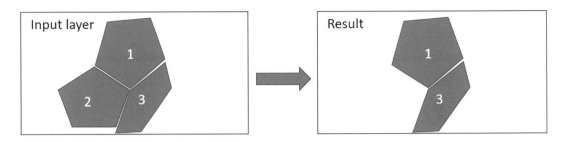

Figure 8-1. *Schematic drawing of the erase operation. Image by author*

In this schematic drawing, you can see that there are three polygons on the left (numbered 1, 2, and 3). The delete operation has deleted polygon 2, which makes that there are only two polygons remaining in the output on the right. Polygon 2 was deleted, or with a synonym, erased. The table containing the data would be affected as shown in Figure 8-2.

ID	Value
1	...
2	...
3	...

ID	Value
1	...
3	...

Figure 8-2. *The table view of this operation. Image by author*

This operation is relatively easy to understand and would not need to be covered in much depth. However, this operation is not what is generally meant when talking about an erase operation in GIS.

The erase function in GIS is actually much closer to the other spatial operations that we have covered before. The erase operation takes two inputs: an input layer, which is the one that we are applying an erase operation on, and the erase features.

The input layer can be any type of vector layer: polygon, line, points, or even mixed. The erase feature generally has to be a polygon, although some implementations may allow you to use other types as well.

What the operation does is erasing all the data in the input layer that are inside the eraser polygon. This will delete a part of your input data, generally because you don't need it.

Let's look at some schematic overviews of erasing on the different input data types in the next sections.

Schematic Overview of Spatially Erasing Points

Spatially erasing points is really quite similar to the standard delete operation. When doing a spatial erase, you are going to delete all parts of the input features that coincide spatially with the erase feature. However, as points do not have a size, they will be either entirely inside or entirely outside of the erase feature, which causes each point to be either entirely deleted or retained. The schematic drawing in Figure 8-3 shows how this works.

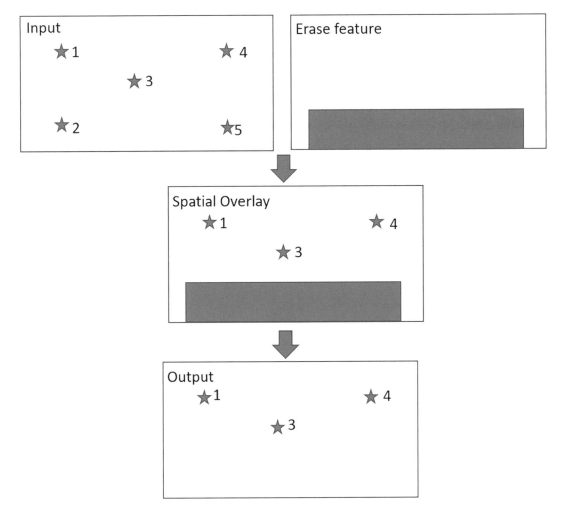

Figure 8-3. *Schematic drawing of spatially erasing points. Image by author*

You can see how the data table would change before and after the operation in the schematic drawing in Figure 8-4.

ID	Value
1	...
2	...
3	...
4	...
5	...

ID	Value
1	...
3	...
4	...

Figure 8-4. *The table view behind the spatial erase. Image by author*

You can see that the features 2 and 5 have simply been removed by the erase operation. This could have been done also using a drop of the features with IDs 2 and 5. Although using a spatial eraser rather than an eraser by ID for deleting a number of points gives the same functional result, it can be very useful and even necessary to use a spatial erase here.

When you have an erase feature, you would not yet have the exact IDs of the points that you want to drop. In this way, the only way to get the list of IDs automatically is to do a spatial join, or an overlay, which is what happens in the spatial erase.

When using more complex features like lines and polygons, the importance of the spatial erase is even larger, as you will see now.

Schematic Overview of Spatially Erasing Lines

Let's now see how this works when doing a spatial erase on line data. In Figure 8-5, you see that there are two line features and a polygon erase feature.

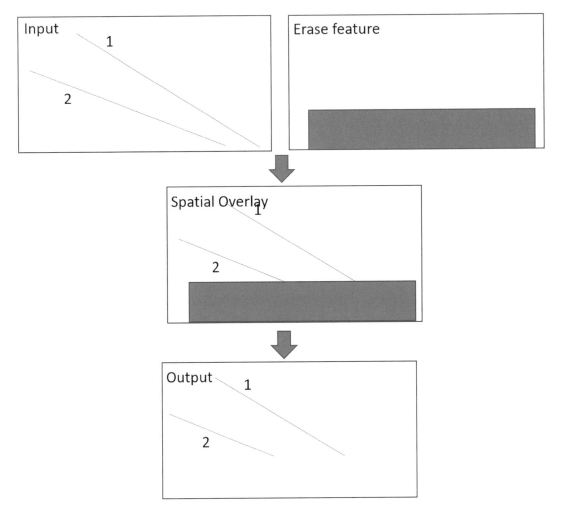

Figure 8-5. *Erasing lines. Image by author*

What happens here is quite different from what happened in the point example. Rather than deleting or keeping entire features, the spatial erase has now made an alteration to the features. Before and after, the data still consists of two lines, yet they are not exactly the same lines anymore. Only a part of each individual feature was erased, thereby not changing the number of features but only the geometry. In the data table, this would look something like shown in Figure 8-6.

ID	Geometry
1	... "Longer line" ...
2	... "Longer line"...

ID	Geometry
1	... "Shorter line" ...
2	... "Shorter line"...

Figure 8-6. *The table view of erasing lines. Image by author*

In the next section, you'll see how this works for polygons.

Schematic Overview of Spatially Erasing Polygons

The functioning of a spatial erase on polygons is essentially the same as the spatial erase applied to lines. The polygons in the input layer, just like lines, can be altered when they are falling partly inside the erase feature. The schematic drawing in Figure 8-7 shows how this works.

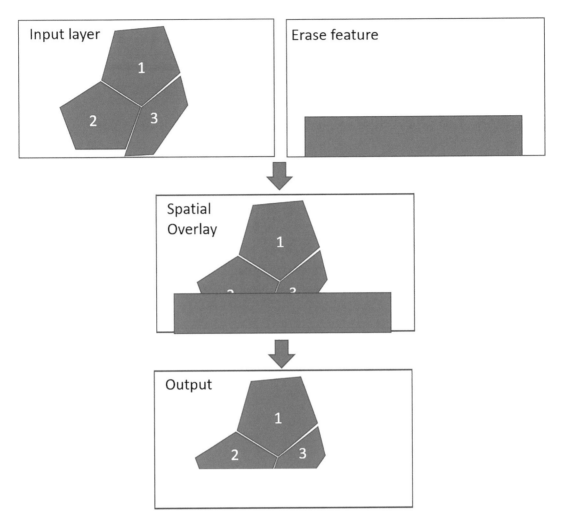

Figure 8-7. *The spatial erase operation applied to polygons. Image by author*

In the drawing, you see that there are three polygons in the input layer on the top left. The erase feature is a rectangular polygon. Using a spatial erase, the output contains altered versions of polygons 2 and 3, since the parts of them that overlaid the erase feature have been cut off.

The impact of this operation in terms of data table would also be similar to the one on the line data. The tables corresponding to this example can be seen in Figure 8-8.

ID	Geometry
1	... "Larger shape" ...
2	... "Larger shape"...
3	... "Larger shape"...

ID	Geometry
1	... "Larger shape" ...
2	... "Smaller shape"...
3	... "Smaller shape"...

Figure 8-8. *The table view of spatially erasing on polygons. Image by author*

You should now have a relatively good intuition about the spatial erase operation. To perfect your understanding, the next section will make an in-depth comparison between the spatial eraser and some comparable operations, before moving on to the implementation in Python.

Erase vs. Other Operations

As you have already seen in the previous three chapters, there are a large number of spatial operations in standard GIS toolkits. This is great, as it allows you to use the tools that are best suitable for your use case. However, it is also essential to look into the details of each of the operations to make sure that you are effectively using the right tool for the task that you have at hand.

Let's now look at a few operations that are relatively comparable to the eraser task. In some cases, you may use another tool for the exact same task. There is no problem in this, as long as you are attentive to the exact implementation and look closely at the documentation of the tools before executing them.

Erase vs. Deleting a Feature

Deleting an entire feature, also called dropping a feature, or dropping a row in your data, was already mentioned earlier as being relatively similar to the spatial eraser. However, there is an essential difference between the spatial eraser and simply dropping an entire feature.

Dropping a feature will only allow you to drop a feature in its entirety. The spatial erase can also do this, but the spatial eraser can do more. When a feature in the input layer is entirely covered by the erase feature, the entire input feature will be deleted, but when the feature in the input layer only overlaps partly with the erase feature, there will be an alteration rather than a deletion. Indeed, the ID of the input feature is not deleted from the dataset, yet its geometry will be altered: reduced by the suppression of the part of the feature that overlapped with the erase feature.

Spatial erase should be used when you want to delete features or part of features based on geometry. Deleting features otherwise is useful when you want to delete based on ID or based on any other column feature of your data.

Erase vs. Clip

You have seen the clipping operation as the first GIS tool that was covered. This tool is made for clipping an input layer to a fixed extent, meaning that you delete everything that is outside of your extent. As clipping is also deleting some part of your data, it may be confused with the spatial erase operation. Let's look at the difference between the two.

The main difference between the clip and the erase operation is that the clip operation will reduce your data to a fixed extent, which can be coordinates or a feature. With a clip, your goal will always be to retain a certain part of the data and delete everything that is outside of your boundaries. In short, you keep inside (where you focus your analysis or map) and remove data that is further away, outside of your boundaries.

The spatial erase does not necessarily take an inside/outside boundaries logic. You can use the eraser also to remove small areas within a general area of interest while keeping the surrounding area.

Whereas both tools are like cookie cutters, the cookie that is cut out is used differently: with the clip operation, you are using the cookie cutter to cut out the cookie that you want to keep, whereas with the erase operation, the cutter is used to cut out parts that you want to throw away. In short, it is a very similar tool for a very different purpose.

Erase vs. Overlay

A third and last operation that is similar to the spatial erase operation is the spatial overlay operation. As described when covering intersections in Chapter 5, there are multiple operations from set theory that can be used on geodata. In that chapter, you mainly focused on the intersection, meaning retaining all data when both layers are spatially intersecting.

Another operation from set theory is the difference operation. You can use the same spatial overlay functionality as for the intersection to compute this difference operation. Rather than retaining parts that intersect in two input layers, the difference operation will keep all features and parts of features from input layer A that are not present in input layer B. Therefore, it will delete from layer A all that is in feature B, similar to an erase.

Depending on the software that you are using, you may or may not encounter implementations of the erase operation. For example, in some versions of the paid ArcGIS software, there is a function to erase. In geopandas, however, there is not, so we might as well use the overlay with the difference operation to obtain the same effect/function/result as the erase operation. This is what we will be doing in the following section.

Erasing in Python

In this exercise, you will be working with a small sample map that was created specifically for this exercise. The data should not be used for any other purpose than the exercise as it isn't very precise, but that is not a problem for now, as the goal here is to master the geospatial analysis tools.

During the coming exercises, you will be working with a mixed dataset of Iberia, which is the peninsula containing Spain and Portugal. The goal of the exercise is to create a map of Spain out of this data, although there is no polygon that indicates the exact region of Spain: this must be created by removing Portugal from Iberia.

I recommend running this code in Kaggle notebooks or in a local environment, as there are some problems in Google Colab for creating overlays. To get started with the exercise, you can import the data using the code in Code Block 8-1.

Code Block 8-1. Importing the data

```
import geopandas as gpd
import fiona

gpd.io.file.fiona.drvsupport.supported_drivers['KML'] = 'rw'
all_data = gpd.read_file('chapter_08_data.kml')
all_data
```

You will obtain a dataframe like the one in Figure 8-9.

	Name	Description	geometry
0	Iberia		POLYGON Z ((-8.99346 37.08769 0.00000, -8.3873...
1	Portugal		POLYGON Z ((-8.89116 41.85535 0.00000, -9.6306...
2	Road 1		LINESTRING Z (-9.17050 38.63791 0.00000, -8.64...
3	Road 2		LINESTRING Z (-8.55244 40.68231 0.00000, -7.18...
4	Road 3		LINESTRING Z (-8.71652 37.12463 0.00000, -7.59...
5	Road 4		LINESTRING Z (-8.63854 41.15617 0.00000, -8.51...
6	Road 5		LINESTRING Z (2.18149 41.29864 0.00000, -0.865...
7	Bilbao		POINT Z (-3.01868 43.30943 0.00000)
8	Barcelona		POINT Z (2.01306 41.50882 0.00000)
9	Madrid		POINT Z (-3.72678 40.39636 0.00000)
10	Sevilla		POINT Z (-6.01832 37.42718 0.00000)
11	Malaga		POINT Z (-4.42530 36.75238 0.00000)
12	Porto		POINT Z (-8.56714 41.19960 0.00000)
13	Lissabon		POINT Z (-9.08350 38.72438 0.00000)
14	Faro		POINT Z (-7.90616 37.11447 0.00000)
15	Santiago de Compostella		POINT Z (-8.54272 42.92122 0.00000)

Figure 8-9. *The data table. Image by author*

Within this dataframe, you can see that there is a mix of data types or geometries. The first two rows contain polygons of Iberia (which is the contour of Spain plus Portugal). Then you have a number of roads, which are lines, followed by a number of cities, which are points.

Let's create a quick map to see what we are working with exactly using Code Block 8-2. You can use the code hereafter to do so. If you are not yet familiar with these methods for plotting, I recommend going back into earlier chapters to get more familiar with this. From here on, we will go a bit faster over the basics of data imports, file formats, data types, and mapping as they have all been extensively covered in earlier parts of this book.

Code Block 8-2. Visualizing the data

```
import contextily as cx

# plotting all data
ax = all_data.plot(markersize=128, figsize=(15,15), edgecolor='black',
facecolor='none')

# adding a contextily basemap
cx.add_basemap(ax, crs=all_data.crs)
```

As a result of this code, you will see the map shown in Figure 8-10.

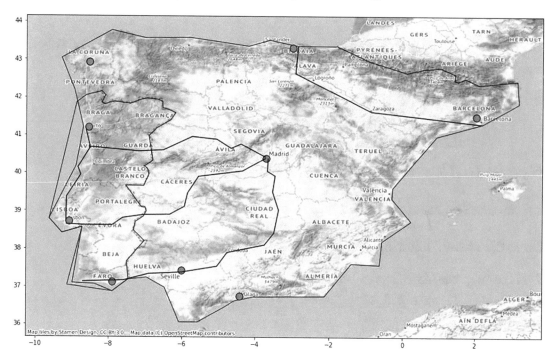

Figure 8-10. *The map resulting from Code Block 8-2. Image by author*

This map is not very clear, as it still contains multiple types of data that are also overlapping. As the goal here is to filter out some data, let's do that first, before working on improved visualizations.

Erasing Portugal from Iberia to Obtain Spain

The goal is to obtain a map of Spain and remove everything that is in Portugal from our data. A first step for this is to create a polygon of Spain. As we know that Iberia consists of Spain and Portugal, we just have to take the difference of Iberia and Portugal to obtain Spain. This is done using the code in Code Block 8-3.

Code Block 8-3. Filter the data based on Name

```
# filter the data based on name
iberia = all_data[all_data['Name'] == 'Iberia']
iberia
```

The resulting dataframe contains only the polygon corresponding to Iberia. This dataframe is shown in Figure 8-11.

	Name	Description	geometry
0	Iberia		POLYGON Z ((-8.99346 37.08769 0.00000, -8.3873...

Figure 8-11. *The data resulting from Code Block 8-3. Image by author*

Now, let's do the same for Portugal, using the code in Code Block 8-4.

Code Block 8-4. Filtering on Portugal

```
# filtering the data based on name
portugal = all_data[all_data['Name'] == 'Portugal']
portugal
```

The resulting dataframe contains only the polygon corresponding to Portugal. This dataframe looks as shown in Figure 8-12.

	Name	Description	geometry
1	Portugal		POLYGON Z ((-8.89116 41.85535 0.00000, -9.6306...

Figure 8-12. *The data resulting from Code Block 8-4. Image by author*

Now, we have two separate objects: one geodataframe containing just the polygon for Iberia and a second geodataframe containing only one polygon for Portugal. What we want to obtain is the difference between the two, as Spain is the part of Iberia that is not Portugal. The code in Code Block 8-5 does exactly this.

Code Block 8-5. Difference operation

```
# overlay to remove portugal from Iberia and create spain
spain = iberia.overlay(portugal, 'difference')
# plot the resulting polygon of spain
spain.plot()
```

The resulting plot is shown in Figure 8-13.

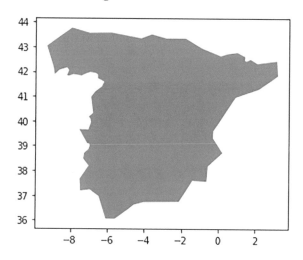

Figure 8-13. *The map resulting from Code Block 8-5. Image by author*

Although the shape is correct, when we look at the spain dataframe object, we can see that the name is still set to the input layer, which is Iberia. You can print the data easily with Code Block 8-6.

Code Block 8-6. Print the data

```
spain
```

The result is shown in Figure 8-14.

	Name	Description	geometry
0	Iberia	POLYGON Z ((-8.94952 42.19360 0.00000, -9.3010...

Figure 8-14. *The Spain data. Image by author*

We can reset the name to Spain using the code in Code Block 8-7.

Code Block 8-7. Setting the name to Spain

```
spain.Name = 'Spain'
spain
```

The resulting dataframe now has the correct value in the column Name, as can be seen in Figure 8-15.

	Name	Description	geometry
0	Spain	POLYGON Z ((-8.94952 42.19360 0.00000, -9.3010...

Figure 8-15. *The Spain data with the correct name. Image by author*

Let's plot all that we have done until here using a background map, so that we can keep on adding to this map of Spain in the following exercises. The code to create this plot is shown in Code Block 8-8.

Code Block 8-8. Plotting the new Spain polygon

```
# plot the new Spain polygon
ax = spain.plot(figsize=(15,15), edgecolor='black', facecolor='none')
# add a contextily basemap
cx.add_basemap(ax, crs=spain.crs)
```

The resulting map is shown in Figure 8-16.

Figure 8-16. *The plot resulting from Code Block 8-8. Image by author*

If you are familiar with the shape of Spain, you will see that it corresponds quite well on this map. We have successfully created a polygon for the country of Spain, just using a spatial operation with two other polygons. You can imagine that such work can occur regularly when working with spatial data, whether it is for spatial analysis, mapping and visualizations, or even for feature engineering in machine learning.

In the following section, you will continue this exercise by also removing the Portuguese cities from our data, so that we only retain relevant cities for our Spanish dataset.

Erasing Points in Portugal from the Dataset

In this second part of the exercise, you will be working on the point data that represents the cities of Iberia. In the dataset, you have some of the major cities of Iberia, meaning that there are some cities in Spain, but some other cities in Portugal. The goal of the exercise is to filter out those cities that are in Spain and remove those that are in Portugal.

In previous chapters, you have seen some techniques that could be useful for this. One could imagine, for example, doing a join with an external table. This external table could be a map from city to country, so that after joining the geodataframe to this lookup table, you could simply do a filter based on country.

In the current exercise, we are taking a different approach, namely, using a spatial overlay with a difference operation, which is the same as an erase. This way, we will erase from the cities all those that have a spatial overlay with the polygon of Portugal.

The first step of this exercise is to create a separate geodataframe that contains only the cities. It is always easier to work with datasets that have one and only one data type. Let's use the code in Code Block 8-9 to filter out all point data, which are the cities in our case.

Code Block 8-9. Filter out all point data

```
from shapely.geometry.point import Point
cities_filter = all_data.geometry.apply(lambda x: type(x) == Point)
cities = all_data[cities_filter]
cities
```

You will obtain the dataset as shown in Figure 8-17, which contains only cities.

	Name Description	geometry
7	Bilbao	POINT Z (-3.01868 43.30943 0.00000)
8	Barcelona	POINT Z (2.01306 41.50882 0.00000)
9	Madrid	POINT Z (-3.72678 40.39636 0.00000)
10	Sevilla	POINT Z (-6.01832 37.42718 0.00000)
11	Malaga	POINT Z (-4.42530 36.75238 0.00000)
12	Porto	POINT Z (-8.56714 41.19960 0.00000)
13	Lissabon	POINT Z (-9.08350 38.72438 0.00000)
14	Faro	POINT Z (-7.90616 37.11447 0.00000)
15	Santiago de Compostella	POINT Z (-8.54272 42.92122 0.00000)

Figure 8-17. *The dataset resulting from Code Block 8-9. Image by author*

Now that we have a dataset with only cities, we still need to filter out the cities of Spain and remove the cities of Portugal. As you can see, there is no other column that we could use to apply this filter, and it would be quite cumbersome to make a manual list of all the cities that are Spanish vs. Portuguese. Even if it would be doable for the current exercise, it would be much more work if we had a larger dataset, so it is not a good practice.

The following code shows how to remove all the cities that have an overlay with the Portugal polygon. Setting the how parameter to "difference" makes that they are removed rather than retained. As a reminder, you have seen other parameters like intersection and union being used in previous chapters. If you don't remember what the other versions do, it would be good to have a quick look back at this point using Code Block 8-10.

Code Block 8-10. Difference operation

```
spanish_cities = cities.overlay(portugal, how = 'difference')
spanish_cities
```

The spanish_cities dataset is shown in Figure 8-18.

	Name Description	geometry
0	Bilbao	POINT Z (-3.01868 43.30943 0.00000)
1	Barcelona	POINT Z (2.01306 41.50882 0.00000)
2	Madrid	POINT Z (-3.72678 40.39636 0.00000)
3	Sevilla	POINT Z (-6.01832 37.42718 0.00000)
4	Malaga	POINT Z (-4.42530 36.75238 0.00000)
5	Santiago de Compostella	POINT Z (-8.54272 42.92122 0.00000)

Figure 8-18. *The dataset resulting from Code Block 8-10. Image by author*

When comparing this with the previous dataset, you can see that indeed a number of cities have been removed. The Spanish cities that are kept are Bilbao, Barcelona, Madrid, Seville, Malaga, and Santiago de Compostela. The cities that are Portuguese have been removed: Porto, Lisbon, and Faro. This was the goal of the exercise, so we can consider it successful.

As a last step, it would be good to add this all to the map that we started to make in the previous section. Let's add the Spanish cities onto the map of the Spanish polygon using the code in Code Block 8-11.

Code Block 8-11. Add the Spanish cities on the map

```
ax = spain.plot(figsize=(15,15), edgecolor='black', facecolor='none')
spanish_cities.plot(ax=ax, markersize=128)
cx.add_basemap(ax, crs=spain.crs)
```

This code will result in the map shown in Figure 8-19, which contains the polygon of the country Spain, the cities of Spain, and a contextily basemap for nicer visualization.

Figure 8-19. *The map resulting from Code Block 8-11. Image by author*

We have now done two parts of the exercise. We have seen how to cut the polygon, and we have filtered out the cities of Spain. The only thing that remains to be done is to resize the roads and make sure to filter out only those parts of the roads that are inside of the Spain polygon. This will be the goal of the next section.

Cutting Lines to Be Only in Spain

In this section, we will work on the line data, which is the only part of the input data that we have not treated yet. The input dataset contains five roads. If you look back at the first overall plot of the data, you can see that some of those roads go through both countries, whereas there are some that are just inside one country. The operation is partly filtering, but also altering, as the multicountry roads need to be cut to the size of the Spanish polygon.

Let's start by creating a dataframe with only the roads, as it is always better to have single-type datasets. This is done in Code Block 8-12.

Code Block 8-12. Create a dataframe with only roads

```
from shapely.geometry.linestring import LineString
roads_filter = all_data.geometry.apply(lambda x: type(x) == LineString)
roads = all_data[roads_filter]
roads
```

You now obtain a dataset that contains only the roads, just like shown in Figure 8-20.

	Name	Description	geometry
2	Road 1		LINESTRING Z (-9.17050 38.63791 0.00000, -8.64...
3	Road 2		LINESTRING Z (-8.55244 40.68231 0.00000, -7.18...
4	Road 3		LINESTRING Z (-8.71652 37.12463 0.00000, -7.59...
5	Road 4		LINESTRING Z (-8.63854 41.15617 0.00000, -8.51...
6	Road 5		LINESTRING Z (2.18149 41.29864 0.00000, -0.865...

Figure 8-20. *The dataset resulting from Code Block 8-12. Image by author*

The problem is not really clear from the data, so let's make a plot to see what is wrong about those LineStrings using the code in Code Block 8-13.

Code Block 8-13. Plot the data

```
ax = spain.plot(figsize=(15,15), edgecolor='black', facecolor='none')
spanish_cities.plot(ax=ax, markersize=128)
roads.plot(ax=ax, linewidth=4, edgecolor='grey')
cx.add_basemap(ax, crs=spain.crs)
```

This code will generate the map in Figure 8-21, which shows that there are a lot of parts of road that are still inside Portugal, which we do not want for our map of Spain.

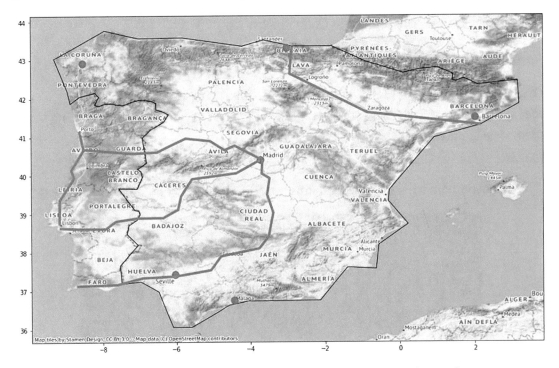

Figure 8-21. *The map resulting from Code Block 8-13. Image by author*

Indeed, you can see here that there is one road (from Porto to Lisbon) that needs to be removed entirely. There are also three roads that start in Madrid and end up in Portugal, so we need to cut off the Portuguese part of those roads.

This is all easy to execute using a difference operation within an overlay again, as is done in the code in Code Block 8-14.

Code Block 8-14. Difference operation

```
spanish_roads = roads.overlay(portugal, how = 'difference')
spanish_roads
```

The result is shown in Figure 8-22.

	Name	Description	geometry
0	Road 1		LINESTRING Z (-7.02586 38.89085 0.00000, -6.34...
1	Road 2		LINESTRING Z (-6.84251 40.58799 0.00000, -6.51...
2	Road 3		LINESTRING Z (-7.46529 37.19099 0.00000, -6.90...
3	Road 5		LINESTRING Z (2.18149 41.29864 0.00000, -0.865...

Figure 8-22. *The data resulting from Code Block 8-14. Image by author*

You can clearly see that some roads are entirely removed from the dataset, because they were entirely inside of Portugal. Roads that were partly in Portugal and partly in Spain were merely altered, whereas roads that were entirely in Spain are kept entirely.

Let's now add the roads to the overall map with the country polygon and the cities, to finish our final map of only Spanish features. This is done in Code Block 8-15.

Code Block 8-15. Add the roads to the overall map

```
ax = spain.plot(figsize=(15,15), edgecolor='black', facecolor='none')
spanish_cities.plot(ax=ax, markersize=128)
spanish_roads.plot(ax=ax, linewidth=4, edgecolor='grey')
cx.add_basemap(ax, crs=spain.crs)
```

The resulting map is shown in Figure 8-23.

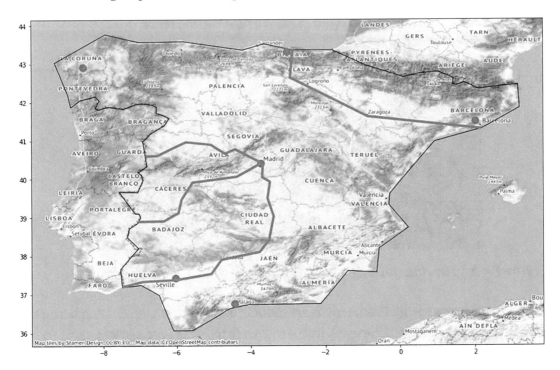

Figure 8-23. *The map resulting from Code Block 8-15. Image by author*

This map shows all of the features reduced to Spain, whereas we started from a dataset in which we did not even have a Spain polygon. These examples show the type of work that is very common in spatial analysis or feature engineering for spatial machine learning. After all, data is not always clean and perfect and often needs some work to be usable.

In this chapter, and the previous chapters, you should have found the basics for working with geospatial data and have enough background to find out how to do some of the other geospatial operations using documentations and other sources. In the next chapters, the focus will shift to more mathematics and statistics, as we will be moving into the chapters on machine learning.

Key Takeaways

1. The erase operation has multiple interpretations. In spatial analysis, its definition is erasing features or parts of features based on a spatial overlay with a specified erase feature.

2. Depending on exact implementation, erasing is basically the same as the difference operation in overlays, which is one of the set theory operations.

3. You can use the difference overlay to erase data from vector datasets (points, lines, or polygons).

4. When erasing on points, you will end up erasing or keeping the entire point, as it is not possible to cut points in multiple parts.

5. When erasing on lines or polygons, you can erase the complete feature if it is entirely overlaying with the erase feature, but if the feature is only partly overlaying, the feature will be altered rather than removed.

PART III

Machine Learning and Mathematics

CHAPTER 9

Interpolation

After having covered the fundamentals of spatial data in the first four chapters of this book, and a number of basic GIS operations in the past four chapters, it is now time to move on to the last four chapters in which you will see a number of statistics and machine learning techniques being applied to spatial data.

This chapter will cover interpolation, which is a good entry into machine learning. The chapter will start by covering definitions and intuitive explanations of interpolation and then move on to some example use cases in Python.

What Is Interpolation?

Interpolation is a task that is relatively intuitive for most people. From a high-level perspective, interpolation means to fill in missing values in a sequence of numbers. For example, let's take the list of numbers:

> 1, 2, 3, 4, ???, 6, 7, 8, 9, 10

Many would easily be able to find that the number 5 should be at the place where the ??? is written. Let's try to understand why this is so easy. If we want to represent this list graphically, we could plot the value against the position (index) in the list, as shown in Figure 9-1.

© Joos Korstanje 2022
J. Korstanje, *Machine Learning on Geographical Data Using Python*,
https://doi.org/10.1007/978-1-4842-8287-8_9

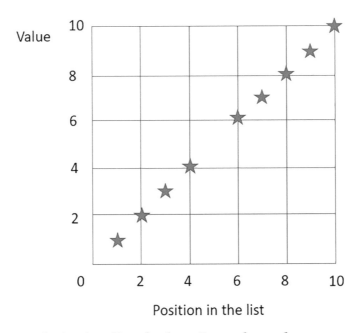

Figure 9-1. *Interpolating in a list of values. Image by author*

When seeing this, we would very easily be inclined to think that this data follows a straight line, as can be seen in Figure 9-2.

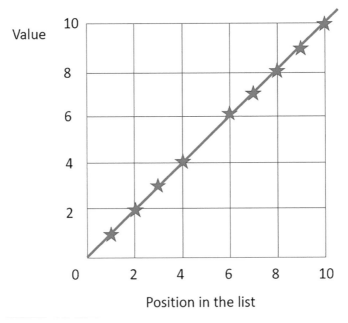

Figure 9-2. *The interpolated line. Image by author*

As we have no idea where these numbers came from, it is hard to say whether this is true or not, but it seems logical to assume that they came from a straight line. Now, let's try another example. To give you a more complex example, try it with the following:

1, ???, 4, ???, 16

If you are able to find it, your most likely guess would be the doubling function, which could be presented graphically as shown in Figure 9-3.

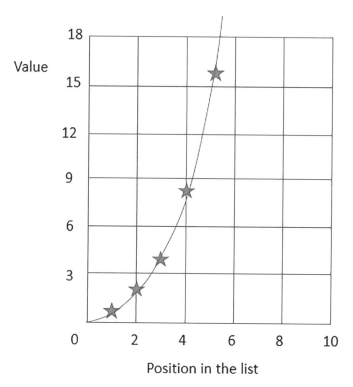

Figure 9-3. Interpolation with a curved line. Image by author

When doing interpolation, we try to find the best estimate for a value in between other values based on a mathematical formula that seems to fit our data. Although interpolation is not necessarily a method in the family of machine learning methods, it is a great way to start discovering the field of machine learning. After all, interpolation is the goal of best guessing some formula to represent data, which is fundamentally what machine learning is about as well. But more on that in the next chapter. Let's first deep dive into a bit of the technical details of how interpolation works and how it can be applied on spatial data.

Different Types of Interpolation

Let's start by covering a number of methods for interpolation that can be used either in "standard" interpolation or in spatial interpolation. The most straightforward method is linear interpolation.

Linear Interpolation

The most straightforward method for interpolation is linear interpolation. Linear interpolation comes down to drawing a straight line from each point to the next and estimating the in-between values to be on that line. The graph in Figure 9-4 shows an example.

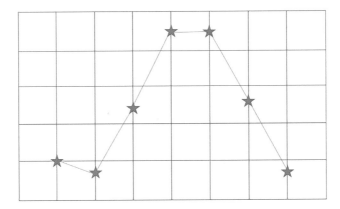

Figure 9-4. *Linear interpolation. Image by author*

Although it seems not such a bad idea, it is not really precise either. The advantage of linear interpolation is that generally it is not very wrong: you do not risk estimating values that are way out of bounds, so it is a good first method to try.

The mathematical function for linear interpolation is the following:

$$y = y_0 + (x - x_0) * \frac{y_1 - y_0}{x_1 - x_0}$$

If you input the value for x at which you want to compute a new y, and the values of x and y of the point before (x_0, y_0) and after (x_1, y_1) your new point, you obtain the new y value of your point.

Polynomial Interpolation

Polynomial interpolation is a bit better for estimating such functions, as polynomial functions can actually be curved. As long as you can find an appropriate polynomial function, you can generally find a relatively good approximation. This could be something like Figure 9-5.

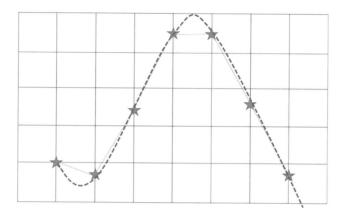

Figure 9-5. *Polynomial interpolation. Image by author*

A risk of polynomial estimation is that it might be very difficult to actually find a polynomial function that fits with your data. If the identified polynomial is highly complex, there is a big risk of having some "crazy curves" somewhere in your function, which will make some of your interpolated values very wrong.

It would be a bit much to go into a full theory of polynomials here, but in short, the formula of a polynomial is any form of a function that contains squared effects, such as

$$f(x) = a + bx + cx^2$$

This can complexify up to very complex forms, for example

$$f(x) = a + bx + cx^2 + dx^3 + ex^4 + gx^5 + hx^6 + ix^7$$

Many, many other forms of polynomials exist. If you are not aware of polynomials, it would be worth it checking out some online resources on the topic.

Piecewise Polynomial or Spline

The piecewise polynomial is an alternative that fits multiple simpler polynomials rather than fitting one polynomial for all the data. This makes that there is a lower risk of making those very wrong estimates that the regular polynomial introduces.

Although piecewise polynomials are easier in their final form, they are not necessarily easier to estimate. Whereas with finding a "regular" polynomial the challenge would be a search of the simplest polynomial that goes at least through all of the points, the problem with our piecewise polynomial is that there are many potential solutions for this.

However, as we fix some rules for ourselves, this problem appears quite easy to solve. In practice, there are good implementations available for fitting things like the cubic spline and other variants. There is no need reinventing the wheel for this topic, as the existing solutions are quite well developed. You could look at the Scipy package in Python for a large availability of such algorithms.

Nearest Neighbor Interpolation

Finally, another common standard method for interpolation is nearest neighbor interpolation. It comes down to estimating a value by finding its closest neighbor and copying their value. The advantage of this method is that you are sure that all the values that you estimate are actually existing at least somewhere in your data, so you are sure that you are not "inventing" anything that is actually impossible. This comes at the cost of making your estimates really unsmooth. The example in Figure 9-6 adds nearest neighbor interpolation to the graph.

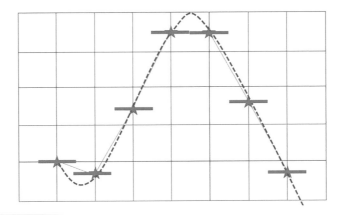

Figure 9-6. *Adding nearest neighbor interpolation to the graph. Image by author*

This nearest neighbor interpolation will assign the value that is the same value as the closest point. The line shape is therefore a piecewise function: when arriving closer (on the x axis) to the next point, the interpolated value (y axis) makes a jump to the next value on the y axis. As you can see, this really isn't the best idea for the curve at hand, but in other situations, it can be a good and easy-to-use interpolation method.

From One-Dimensional to Spatial Interpolation

Now that you have seen some intuitive fundamentals of interpolation, it is time to move on to spatial interpolation. Of course, the goal of this chapter is to talk about interpolation of spatial data, and it may be hard to imagine how we move from the previous graphs to an interpolation in the two-dimensional case of spatial interpolation. Or you can imagine the previous graphs are side views of spatial interpolated values when you stand on the ground.

As visuals are generally a great way to understand interpolation methods, let's start discussing spatial interpolation with a very common use case: interpolation temperature over a country that has only a few numbers of official national thermometers. Imagine that you have a rectangular country on the map in Figure 9-7, with the temperature measures taken at each point.

Figure 9-7. An interpolation exercise. Image by author

Depending on where we live, we want to have the best appropriate value for ourselves. In the north of the country, it is 10 degrees Celsius; in the south, it is 0 degree Celsius. Let's use a linear interpolation for this, with the result shown in Figure 9-8.

10	10	10	10	10	10	10	10
7.5	7.5	7.5	7.5	7.5	7.5	7.5	7.5
5	5	5	5	5	5	5	5
2.5	2.5	2.5	2.5	2.5	2.5	2.5	2.5
0	0	0	0	0	0	0	0

Figure 9-8. *The result of the exercise using linear interpolation. Image by author*

This linear approach does not look too bad, and it is easy to compute by hand for this data. Let's also see what would have happened with a nearest neighbor interpolation, which is also easy to do by hand. It is shown in Figure 9-9.

10	10	10	10	10	10	10	10
10	10	10	10	10	10	10	10
0	0	0	0	0	0	0	0
0	0	0	0	0	0	0	0

Figure 9-9. *The result of the exercise using nearest neighbor interpolation. Image by author*

The middle part has been left out, as defining ties is not that simple, yet you can get the idea of what would have happened with a nearest neighbor interpolation in this example.

For the moment, we will not go deeper into the mathematical definitions, but if you want to go deeper, you will find many resources online. For example, you could get started here: `https://towardsdatascience.com/polynomial-interpolation-3463ea4b63dd`. For now, we will focus on applications to geodata in Python.

Spatial Interpolation in Python

In this section, you will see how to apply some of these interpolation methods in Python. Let's start by generating some example data points that we can apply this interpolation to. Imagine we have only data on the following four points. Each data point is a temperature measurement at the same timestamp, but at a different location. The goal of the exercise is to use interpolation to estimate these values at unmeasured, in-between locations. The data looks as shown in Code Block 9-1.

Code Block 9-1. The data

```
data = { 'point1': {
            'lat': 0,
            'long': 0,
            'temp': 0 },
        'point2': {
          'lat': 10,
          'long': 10,
          'temp': 20 },
        'point3' : {
          'lat': 0,
          'long': 10,
          'temp': 10 },
        'point4': {
          'lat': 10,
          'long': 0,
          'temp': 30 }
}
```

Now, the first thing that we can do is making a dataframe from this dictionary and getting this data into a geodataframe. For this, the easiest is to make a regular pandas dataframe first, using the code in Code Block 9-2.

Code Block 9-2. Creating a dataframe

```
import pandas as pd
df = pd.DataFrame.from_dict(data, orient='index')
df
```

We see the dataframe being successfully created in Figure 9-10.

	lat	long	temp
point1	0	0	0
point2	10	10	20
point3	0	10	10
point4	10	0	30

Figure 9-10. *The data imported in Python. Image by author*

As a next step, let's convert this dataframe into a geopandas geodataframe, while specifying the geometry to be point data, with latitude and longitude. This is done in Code Block 9-3.

Code Block 9-3. Converting to geodataframe

```
import geopandas as gpd
from shapely.geometry.point import Point
gdf = gpd.GeoDataFrame(df, geometry=[Point(y,x) for x,y in
zip(list(df['lat']), list(df['long']))])
gdf.plot(markersize=(gdf['temp']+1)*10)
```

The plot that results from this code is shown in Figure 9-11.

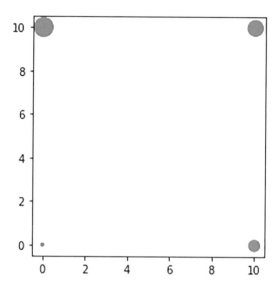

Figure 9-11. *The plot resulting from Code Block 9-3. Image by author*

As you can see in this plot, there are four points with a different size. From a high-level perspective, it seems quite doable to find intermediate values to fill in between the points. What is needed to do so, however, is to find a mathematical formula in Python that represents this interpolation and then use it to predict the interpolated values.

Linear Interpolation Using Scipy Interp2d

The first thing that we are going to try is a linear 2D interpolation. For this, we use Scipy Interp2d, which is a function that returns an interpolation function. Once this function has been created using the input data, you can actually call it to make predictions for any new point.

The code in Code Block 9-4 shows how to create the 2D linear interpolation function using the four input points.

Code Block 9-4. Creating an interpolation function

```
from scipy.interpolate import interp2d
my_interpolation_function = interp2d(df['lat'], df['long'], df['temp'],
kind='linear')
```

Now that we have this function, we can call it on new points. However, we first need to define which points we are going to use for the interpolation. As we have four points in a square organization, let's interpolate at the point exactly in the middle and the points that are in the middle along the sides. We can create this new df using the code in Code Block 9-5.

Code Block 9-5. The new dataframe

```
new_points = {'point5': {'lat': 0, 'long': 5},
              'point6': {'lat': 5, 'long': 0},
              'point7': {'lat': 5, 'long': 5},
              'point8': {'lat': 5, 'long': 10},
              'point9': {'lat': 10, 'long': 5}
              }
new_df = pd.DataFrame.from_dict(new_points, orient='index')
new_df
```

The resulting dataframe is shown in Figure 9-12.

	lat	long
point5	0	5
point6	5	0
point7	5	5
point8	5	10
point9	10	5

Figure 9-12. *The resulting dataframe. Image by author*

The data here only has latitude and longitude, but it does not yet have the estimated temperature. After all, the goal is to use our interpolation function to obtain these estimated temperatures.

In Code Block 9-6, you can see how to loop through the new points and call the interpolation function to estimate the temperature on this location. Keep in mind that this interpolation function is the mathematical definition of a linear interpolation, based on the input data that we have given.

Code Block 9-6. Applying the interpolation

```
interpolated_temps = []
for i,row in new_df.iterrows():
  interpolated_temps.append(my_interpolation_function(row['lat'],
  row['long'])[0])

new_df['temp'] = interpolated_temps
new_df
```

You can see the numerical estimations of these results in Figure 9-13.

	lat	long	temp
point5	0	5	5.0
point6	5	0	15.0
point7	5	5	15.0
point8	5	10	15.0
point9	10	5	25.0

Figure 9-13. *The estimations resulting from Code Block 9-6. Image by author*

The linear interpolation is the most straightforward, and it is clear from those predictions that they look solid. It would be hard to say whether they are good or not, as we do not have any ground truth value in interpolation use cases, yet we can say that it is nothing too weird at least.

Now that we have estimated them, we should try to do some sort of analysis. Combining them into a dataframe with everything so that we can rebuild the plot is done in Code Block 9-7.

Code Block 9-7. Put the data together and plot

```
combi_df = pd.concat([df, new_df])
gdf = gpd.GeoDataFrame(combi_df, geometry=[Point(y,x) for x,y in
zip(list(combi_df['lat']), list(combi_df['long']))])
gdf.plot(markersize=(gdf['temp']+1)*10)
```

This code will output the plot shown in Figure 9-14.

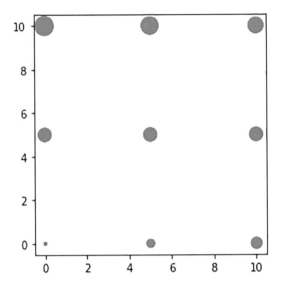

Figure 9-14. *The plot with interpolated values. Image by author*

Even though we do not have an exact metric to say whether this interpolation is good or bad, we can at least say that the interpolation seems more or less logical to the eye, which is comforting. This first try appears rather successful. Let's try out some more advanced methods in the next section, to see how results may differ with different methods.

Kriging

In the first part of this chapter, you have discovered some basic, fundamental approaches to interpolation. The thing about interpolation is that you can make it as simple, or as complex, as you want.

Although the fundamental approaches discussed earlier are often satisfactory in practical results and use cases, there are some much more advanced techniques that we need to cover as well.

In this second part of the chapter, we will look at Kriging for an interpolation method. Kriging is a much more advanced mathematical definition for interpolation. Although it would surpass the level of this book to go into too much mathematical detail here, for those readers that are at ease with more mathematical details, feel free to check out some online resources like https://en.wikipedia.org/wiki/Kriging and www.publichealth.columbia.edu/research/population-health-methods/kriging-interpolation.

Linear Ordinary Kriging

There are many forms of Kriging. As often with mathematical methods, the more complex they are, the more tuning and settings there are to be done. In the coming sections, let's look at a number of Kriging solutions that would be in competition with the linear interpolation solution given earlier. At the end, we will compare and conclude on the exercise.

To start, you can use the pykrige library with the Ordinary Kriging functionality, as shown in Code Block 9-8. As before, you estimate the function first and then estimate on the new_df. This is done in Code Block 9-8.

Code Block 9-8. Interpolate with Linear Ordinary Kriging

```
from pykrige.ok import OrdinaryKriging
my_ok = OrdinaryKriging(df['long'], df['lat'], df['temp'])
zvalues, sigmasq = my_ok.execute('points', new_df['long'].map(float).
values, new_df['lat'].map(float).values)
new_df['temp'] = zvalues.data
new_df
```

You can see the estimated results in Figure 9-15.

	lat	long	temp
point5	0	5	7.53967
point6	5	0	15.00000
point7	5	5	15.00000
point8	5	10	15.00000
point9	10	5	22.46033

Figure 9-15. *The interpolated values with Linear Ordinary Kriging. Image by author*

Interestingly, some of these estimated values are not the same at all. Let's plot them to see whether there is anything weird or different going on in the plot, using the code in Code Block 9-9.

Code Block 9-9. Plotting the interpolation

```
combi_df = pd.concat([df, new_df])
gdf = gpd.GeoDataFrame(combi_df, geometry=[Point(y,x) for x,y in
zip(list(combi_df['lat']), list(combi_df['long']))])
gdf.plot(markersize=(gdf['temp']+1)*10)
```

The resulting plot is shown in Figure 9-16.

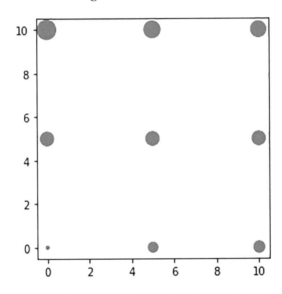

Figure 9-16. *The resulting plot using Linear Ordinary Kriging. Image by author*

There is nothing too wrongly estimated if we judge by the plot, so there is no reason to discount these results. As we have no metric for good or wrong interpolation, this must be seen as just an alternative estimation. Let's try to see what happens when using other settings to Kriging in the next section.

Gaussian Ordinary Kriging

In this section, let's try to change the interpolation model behind the Kriging interpolation. We can tweak the settings by setting the variogram_model parameter to another value. In this case, let's choose "gaussian" to see what happens. This is done in the code in Code Block 9-10.

Code Block 9-10. Gaussian Ordinary Kriging

```
my_ok = OrdinaryKriging(df['long'], df['lat'], df['temp'], variogram_model
= 'gaussian')

zvalues, sigmasq = my_ok.execute('points', new_df['long'].map(float).
values, new_df['lat'].map(float).values)
new_df['temp'] = zvalues.data
new_df
```

Now, the results are given in the table in Figure 9-17.

	lat	long	temp
point5	0	5	3.786315
point6	5	0	15.000000
point7	5	5	15.000000
point8	5	10	15.000000
point9	10	5	26.213685

Figure 9-17. *The result with Gaussian Ordinary Kriging. Image by author*

Interestingly, the estimates for point5 and point9 change quite drastically again! Let's make a plot again to see if anything weird is occurring during this interpolation. This is done in Code Block 9-11.

Code Block 9-11. Plotting the result

```
combi_df = pd.concat([df, new_df])
gdf = gpd.GeoDataFrame(combi_df, geometry=[Point(y,x) for x,y in
zip(list(combi_df['lat']), list(combi_df['long']))])
gdf.plot(markersize=(gdf['temp']+1)*10)
```

The resulting plot is shown in Figure 9-18.

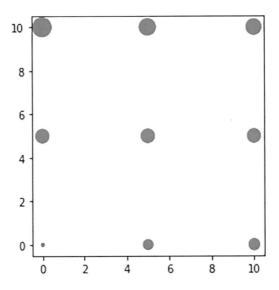

Figure 9-18. *The result from Gaussian Ordinary Kriging. Image by author*

Again, when looking at this plot, it cannot be said that this interpolation is wrong in any way. It is different from the others, but just as valid.

Exponential Ordinary Kriging

As a final test and exercise, let's try again another setting for the variogram. This time, let's use the "exponential" setting to the variogram_model. The code for this is shown in Code Block 9-12.

Code Block 9-12. Exponential Ordinary Kriging

```
my_ok = OrdinaryKriging(df['long'], df['lat'], df['temp'], variogram_model
= 'exponential')

zvalues, sigmasq = my_ok.execute('points', new_df['long'].map(float).
values, new_df['lat'].map(float).values)
new_df['temp'] = zvalues.data
new_df
```

The results of this interpolation are shown in Figure 9-19.

	lat	long	temp
point5	0	5	9.676856
point6	5	0	15.000000
point7	5	5	15.000000
point8	5	10	15.000000
point9	10	5	20.323144

Figure 9-19. *The result from Exponential Ordinary Kriging. Image by author*

Interestingly, again point5 and point9 are the ones that change a lot, while the others stay the same. For coherence, let's make the plot of this interpolation as well, using Code Block 9-13.

Code Block 9-13. Plot the Exponential Ordinary Kriging

```
combi_df = pd.concat([df, new_df])
gdf = gpd.GeoDataFrame(combi_df, geometry=[Point(y,x) for x,y in
zip(list(combi_df['lat']), list(combi_df['long']))])
gdf.plot(markersize=(gdf['temp']+1)*10)
```

The resulting plot is shown in Figure 9-20.

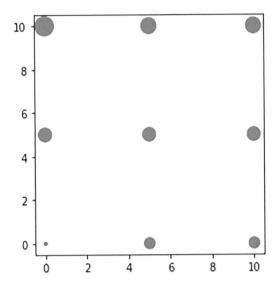

Figure 9-20. *The plot from Exponential Ordinary Kriging. Image by author*

Again, nothing obvious wrong with this plot, yet its results are again different than before. It would only make sense to wonder which of them is right. Let's conclude on this in the next section.

Conclusion on Interpolation Methods

Interestingly, we have started this chapter with a fairly simple interpolation use case. While one might be inclined to think that this question is super easy to solve and can even be solved manually, the variety of possible answers given by the different methods shows that this is not necessarily a trivial task.

To recap, we have obtained all possible answers in Table 9-1.

Table 9-1. *The Results from the Interpolation Benchmark*

	Linear	Linear Kriging	Gaussian Kriging	Exponential Kriging
Point5	5	7.54	3.79	9.68
Point6	15	15	15	15
Point7	15	15	15	15
Point8	15	15	15	15
Point9	25	22.46	26.21	20.32

Even for such a simple interpolation example, we see spectacularly large differences in the estimations of points 5 (middle bottom in the graph) and 9 (right middle in the graph).

Now the big question here is of course whether we can say that any of those are better than the others. Unfortunately, when applying mathematical models to data where there is no ground truth, you just don't know. You can build models that are useful to your use case, you can use human and business logic to assess different estimates, and you can use rules of thumb like Occam's razor (keep the simplest possible model) for your decision to retain one model over the other.

Alternatively, you can also turn to supervised machine learning for this. Classification and regression will be covered in the coming two chapters, and they are also methods for estimating data points that we don't know, yet they are focused much more on performance metrics to evaluate the fit of our data to reality, which is often missing in interpolation use cases.

In conclusion, although there is not necessarily only one good answer, it is always useful to have a basic working knowledge of interpolation. Especially in spatial use cases, it is often necessary to convert data measured at specific points (like temperature stations and much more) into a more continuous view over a larger two-dimensional surface (like countries, regions, and the like). You have seen in this chapter that relatively simple interpolations are already quite efficient in some use cases and that there is a vast complexity to be discovered for those who wanted to go in more depth.

Key Takeaways

1. Interpolation is the task of estimating unknown values in between a number of known values, which comes down to estimating values on unmeasured locations.

2. We generally define a mathematical function or formula based on the known values and then use this function to estimate the values that we do not know.

3. There are many mathematical "base" formulas that you can apply to your points, and depending on the formula you chose, you may end up with quite different results.

4. When interpolating, we generally strive to obtain estimates for points of which we do not have a ground truth value, that is, we really don't know which value is wrong or correct. Cross-validation and other evaluation methods can be used and will be covered in the coming chapters on machine learning.

5. In the case where multiple interpolation methods give different results, we often need to define a choice based on common sense, business logic, or domain knowledge.

CHAPTER 10

Classification

With the current chapter, you are now arriving at one of the main parts of the book about machine learning, namely, classification. Classification is, next to regression and clustering, one of the three main tasks in machine learning, and they will all be covered in this book.

Machine learning is a very large topic, and it would be impossible to cover all of machine learning in just these three chapters. The choice has been made to do a focus on applying machine learning models to spatial data. The focus is therefore on presenting interesting and realizing use cases for machine learning on spatial data while showing how spatial data can be used as an added value with respect to regular data.

There will not be very advanced mathematical, statistical, nor algorithmic discussions in the chapters. There are many standard resources out there for those readers who want to gain a deep and thorough mathematical understanding of machine learning in general.

The chapter will start with a general introduction of what classification is, what we can use it for, and some models and tools that you'll need for doing classification, and then we'll dive into a deep spatial classification use case for the remainder of the chapter. Let's now start with some definitions and introductions first.

Quick Intro to Machine Learning

For those of you who are not very familiar with machine learning, let's do a very brief intro into the matter. In very short, the goal of machine learning is to find patterns in data by creating models that can be reused on new data. There are tons of different models, which are basically mathematical formulas, or algorithms, that are able to identify these patterns in some way or another.

© Joos Korstanje 2022
J. Korstanje, *Machine Learning on Geographical Data Using Python*,
https://doi.org/10.1007/978-1-4842-8287-8_10

A big distinction within machine learning is that of supervised vs. unsupervised models. Unsupervised models are those machine learning models that do find patterns within data, but do not have a ground truth to measure the performance of them, much like what you have seen in the previous chapter on interpolation. Two main categories in unsupervised models are clustering, in which you try to find groups based on the rows of your dataset, and feature reduction, in which you try to redefine your columns in a more optimal way, by analyzing correlations and other relations between them.

Supervised models have a very different take, as there is a real, true value to be predicted here. In supervised models, we have a target variable, which is what we want to predict, and a number of predictor features, also called independent variables or X variables. There are again a lot of specific models that try to best capture mathematical information to determine a predictor function that takes X variables to predict the target.

Quick Intro to Classification

The targets in supervised machine learning are, in most cases, one of two types. If the target is a numeric variable, we are going to be in a case of regression. If the target is a categorical variable, we are in a case of classification.

Without going into the details, you can intuitively understand that predicting a number is a bit different from predicting a category. The fields are related, and the models are often built on the same basis, but the adaptations that are necessary to adapt to the target outcome make that they are considered as two different use cases.

Spatial Classification Use Case

In this chapter, we'll do a use case of classification. Therefore, we are in the supervised machine learning category, so this means that we know a ground truth for our target variable. Also, as we are in classification, the target variable will be categorical.

The use case of this chapter will present GPS tracking data of 20 hypothetical clients of a big mall. The 20 clients have been asked to install a tracking app on their phone, and their GPS coordinates have been collected throughout their mall visit.

It was communicated to the participants that they receive a gift coupon for a 50% discount on a new restaurant. This was done not just to incentivize participants to take part in the study, but it was done also with an ulterior motive. Actually, there is a tracking in place to measure whether the user has used the coupon, so that we can study the link between movement patterns and the potential interest in this new restaurant.

Our classification case will be executed as a marketing study for the new restaurant. The goal of the study is to use a tracked GPS path to predict whether a person is interested in this restaurant. The model, if successful, will be used to implement push notification ads to incentivize clients to find out about this new restaurant.

Sending an ad to a user costs money, so there is a real interest in finding out to which client we want to send this ad or not. We need to build a model that is as good as possible in predicting interest in using the coupon based on only the sequence of GPS points.

Feature Engineering with Additional Data

In this use case, there is a heavy weight of feature engineering, which is done using spatial analysis operations just like those that were covered in the chapter before this. This shows how the topics are related and how spatial operations have a great added value for machine learning on spatial data.

The reason for this is that mere GPS coordinates do not hold much value to predict anything at all. We need to work with this data and transcribe it into variables that can actually be used to create a prediction for coupon interest. Let's move on to have a look at the data, so that it becomes clearer what we are working with. For this use case, it is recommended to use Kaggle notebook environments or any other environment of your choice.

Importing and Inspecting the Data

Let's get started by importing the data. You can use the code in Code Block 10-1 to do so.

Code Block 10-1. Importing the data

```
import geopandas as gpd
import fiona
gpd.io.file.fiona.drvsupport.supported_drivers['KML'] = 'rw'
all_data = gpd.read_file('chapter_10_data.kml')
all_data
```

When looking at this data, you will see something like Figure 10-1.

	Name	Description	geometry
0	Mall	POLYGON Z ((2.77385 48.85390 0.00000, 2.77872 …
1	Restaurants	POLYGON Z ((2.78255 48.85498 0.00000, 2.78317 …
2	High Fashion	POLYGON Z ((2.78239 48.85227 0.00000, 2.78252 …
3	Supermarket	POLYGON Z ((2.77960 48.85517 0.00000, 2.78097 …
4	Cheap Fashion	POLYGON Z ((2.78112 48.85586 0.00000, 2.78102 …
5	Cosmetics	POLYGON Z ((2.78094 48.85481 0.00000, 2.78254 …
6	Electronics	POLYGON Z ((2.78245 48.85536 0.00000, 2.78254 …
7	West Wing	POLYGON Z ((2.77567 48.85444 0.00000, 2.77942 …
8	Bought Yes	LINESTRING Z (2.78016 48.85490 0.00000, 2.7801…
9	Bought Yes	LINESTRING Z (2.78016 48.85490 0.00000, 2.7790…
10	Bought Yes	LINESTRING Z (2.78001 48.85483 0.00000, 2.7790…
11	Bought Yes	LINESTRING Z (2.78019 48.85487 0.00000, 2.7808…
12	Bought Yes	LINESTRING Z (2.78012 48.85483 0.00000, 2.7797…
13	Bought Yes	LINESTRING Z (2.78001 48.85483 0.00000, 2.7787…
14	Bought Yes	LINESTRING Z (2.78054 48.85489 0.00000, 2.7806…
15	Bought Yes	LINESTRING Z (2.78028 48.85486 0.00000, 2.7791…
16	Bought Yes	LINESTRING Z (2.78033 48.85491 0.00000, 2.7812…
17	Bought Yes	LINESTRING Z (2.78021 48.85484 0.00000, 2.7809…
18	Bought No	LINESTRING Z (2.78006 48.85483 0.00000, 2.7805…
19	Bought No	LINESTRING Z (2.78034 48.85489 0.00000, 2.7810…
20	Bought No	LINESTRING Z (2.78001 48.85483 0.00000, 2.7808…
21	Bought No	LINESTRING Z (2.78033 48.85491 0.00000, 2.7811…
22	Bought No	LINESTRING Z (2.78016 48.85490 0.00000, 2.7822…

Figure 10-1. *The data resulting from Code Block 10-1. Image by author*

The dataset is a bit more complex than what we have worked with in previous chapters, so let's make sure to have a good understanding of what we are working with.

The first row of the geodataframe contains an object called the mall. This polygon is the one that covers the entire area of the mall, which is the extent of our study. It is here just for informative purposes, and we won't need it during the exercise.

The following features from rows 1 to 7 present areas of the mall. They are also polygons. Each area can either be one shop, a group of shops, a whole wing, or whatnot, but they generally regroup a certain type of store. We will be able to use this information for our model.

The remaining data are 20 itineraries. Each itinerary is represented as a LineString, that is, a line, which is just a sequence of points that has been followed by each of the 20 participants in the study. The name of each of the LineStrings is either Bought Yes, meaning that they have used the coupon after the study (indicating the product interests them), or Bought No, indicating that the coupon was not used and therefore that the client is probably not interested in the product.

Let's now move on to make a combined plot of all this data to get an even better feel of what we are working with. This can be done using Code Block 10-2.

Code Block 10-2. Plotting the data

```
all_data.plot(figsize=(15,15),alpha=0.1)
```

When executing this code, you will end up with the map in Figure 10-2. Of course, it is not the most visual map, but the goal is here to put everything together in a quick image to see what is going on in the data.

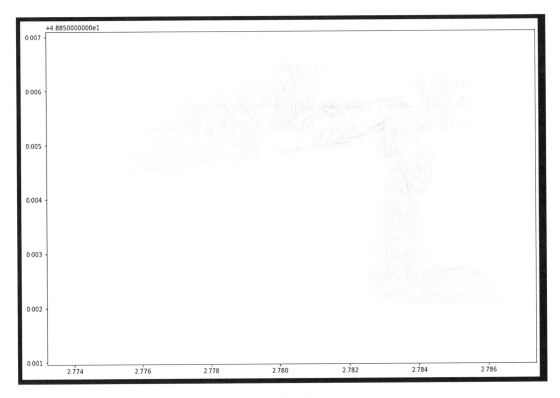

Figure 10-2. *The map resulting from Code Block 10-2*

In this map, the most outer light-gray contours are the contours of the large polygon that sets the total mall area. Within this, you see a number of smaller polygons, which indicate the areas of interest for our study, which all have a specific group of store types inside them. Finally, you also see the lines criss-crossing, which represents the 20 participants of the study making their movements throughout the mall during their visit.

What we want to do now is to use the information of the store segment polygons to annotate the trips of each participant. It would be great to end up with a percentage of time that each participant has spent in each type of store, so that we can build a model that learns a relationship between the types of stores that were visited in the mall and the potential interests in the new restaurant.

As a first step toward this model, let's separate the data to obtain datasets with only one data type. For this, we will need to separate the information polygons from the participant itineraries. Using all that you have seen earlier in the book, that should not be too hard. The code in Code Block 10-3 shows how to get the info polygons in a new dataset.

Code Block 10-3. Select the info polygons into a separate dataset

```
info_polygons = all_data.loc[1:7,:]
info_polygons
```

The result is the dataset with info_polygons shown in Figure 10-3.

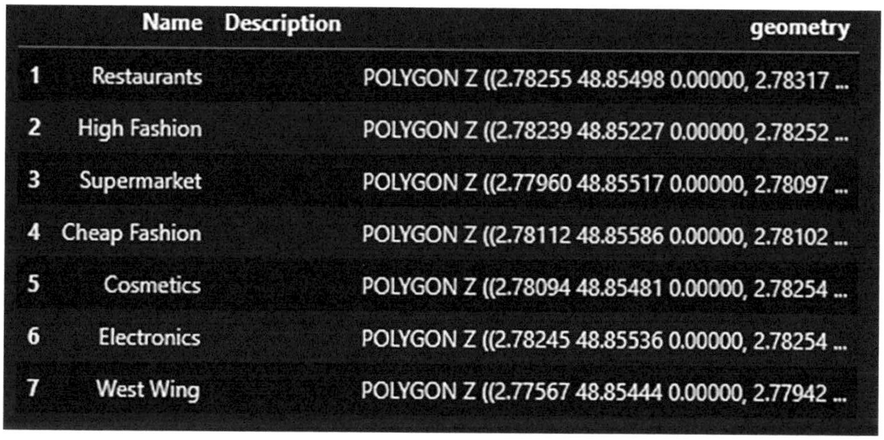

	Name	Description	geometry
1	Restaurants		POLYGON Z ((2.78255 48.85498 0.00000, 2.78317 ...
2	High Fashion		POLYGON Z ((2.78239 48.85227 0.00000, 2.78252 ...
3	Supermarket		POLYGON Z ((2.77960 48.85517 0.00000, 2.78097 ...
4	Cheap Fashion		POLYGON Z ((2.78112 48.85586 0.00000, 2.78102 ...
5	Cosmetics		POLYGON Z ((2.78094 48.85481 0.00000, 2.78254 ...
6	Electronics		POLYGON Z ((2.78245 48.85536 0.00000, 2.78254 ...
7	West Wing		POLYGON Z ((2.77567 48.85444 0.00000, 2.77942 ...

Figure 10-3. *The data resulting from Code Block 10-3. Image by author*

Let's extract the itineraries as well, using the code in Code Block 10-4.

Code Block 10-4. Selecting the itineraries

```
itineraries = all_data.loc[8:,:]
itineraries
```

The itineraries dataset is a bit longer, but it looks as shown in Figure 10-4 (truncated version).

	Name	Description	geometry
8	Bought Yes		LINESTRING Z (2.78016 48.85490 0.00000, 2.7801...
9	Bought Yes		LINESTRING Z (2.78016 48.85490 0.00000, 2.7790...
10	Bought Yes		LINESTRING Z (2.78001 48.85483 0.00000, 2.7790...
11	Bought Yes		LINESTRING Z (2.78019 48.85487 0.00000, 2.7808...
12	Bought Yes		LINESTRING Z (2.78012 48.85483 0.00000, 2.7797...
13	Bought Yes		LINESTRING Z (2.78001 48.85483 0.00000, 2.7787...
14	Bought Yes		LINESTRING Z (2.78054 48.85489 0.00000, 2.7806...
15	Bought Yes		LINESTRING Z (2.78028 48.85486 0.00000, 2.7791...
16	Bought Yes		LINESTRING Z (2.78033 48.85491 0.00000, 2.7812...
17	Bought Yes		LINESTRING Z (2.78021 48.85484 0.00000, 2.7809...
18	Bought No		LINESTRING Z (2.78006 48.85483 0.00000, 2.7805...
19	Bought No		LINESTRING Z (2.78034 48.85489 0.00000, 2.7810...
20	Bought No		LINESTRING Z (2.78001 48.85483 0.00000, 2.7808...
21	Bought No		LINESTRING Z (2.78033 48.85491 0.00000, 2.7811...
22	Bought No		LINESTRING Z (2.78016 48.85490 0.00000, 2.7822...
23	Bought No		LINESTRING Z (2.78003 48.85490 0.00000, 2.7816...
24	Bought No		LINESTRING Z (2.78016 48.85490 0.00000, 2.7790...
25	Bought No		LINESTRING Z (2.78023 48.85493 0.00000, 2.7817...
26	Bought No		LINESTRING Z (2.78054 48.85489 0.00000, 2.7806...
27	Bought No		LINESTRING Z (2.78012 48.85483 0.00000, 2.7811...

Figure 10-4. *Truncated version of the data from Code Block 10-4*

Spatial Operations for Feature Engineering

Now that we have good overview of this dataset, it is time to start doing the spatial operation needed to add the information of participants' presence in each of the locations. For this, we are going to need some sort of spatial overlay operation, as you have seen in the earlier parts of this book.

To do the overlay part, it is easier to do this point by point, as we have lines that are passing through many of the interest areas. If we were to do the operation using lines, we would end up needing to cut the lines according to the boundaries of the polygons and use line length for estimating time spent in the store section.

If we cut the lines into points, we can do a spatial operation to find the presence of each point and then simply count, for each participant, the number of points in each store section. If the points are collected on the basis of equal frequency, the number of points is an exact representation of time.

The reason that we can do without the line is that we do not care about direction or order here. After all, if we wanted to study the order of visits to each store section, we would need to keep information about the sequence. The line is able to do this, whereas the point data type is not.

In this section, you can see what a great advantage it is to master Python for working on geodata use cases. Indeed, it is very advantageous of Python that we have liberty to convert geometry objects to strings and do loops through them which would potentially be way more complex in more click button, precoded, GIS tools.

The disadvantage may be that it is a little hard to get your head around sometimes, but the code in Code Block 10-5 walks you through an approach to get the data from a wide data format (one line per client) to a long data format (one row per data point/coordinate).

Code Block 10-5. Get the data from a wide data format to a long data format

```
import pandas as pd
from shapely.geometry.point import Point

results = []

# split the lines into points
for i, row in itineraries.iterrows():
```

```
    # making the line string into a list of the coordinates as strings and
    removing redundant information
    list_of_points_extracted = str(row['geometry']).strip('LINESTRING Z
    (').strip(')').split(',')
    list_of_points_extracted = [point[:-2] for point in list_of_points_
    extracted]

    # convert lat and long into floats
    list_of_points_extracted = [Point([ float(y) for y in x.strip(' ').
    split(' ')]) for x in list_of_points_extracted]
    list_of_points_extracted = [[i, row.Name] + [x] for x in list_of_
    points_extracted]
    results += list_of_points_extracted

results_df = pd.DataFrame(results)
results_df.columns = ['client_id', 'target', 'point']
results_df
```

The result of this code is the data in a long format: one row per point instead of one row per participant. A part of the data is shown in Figure 10-5.

	client_id	target	point
0	8	Bought Yes	POINT (2.780162 48.8549039)
1	8	Bought Yes	POINT (2.7801834 48.8550522)
2	8	Bought Yes	POINT (2.7801191 48.8552639)
3	8	Bought Yes	POINT (2.7802693 48.8553275)
4	8	Bought Yes	POINT (2.7799367 48.8553275)
...	...	...	...
354	27	Bought No	POINT (2.7801213 48.8548265)
355	27	Bought No	POINT (2.7811352 48.8555854)
356	27	Bought No	POINT (2.7811567 48.8554936)
357	27	Bought No	POINT (2.7813069 48.8554936)
358	27	Bought No	POINT (2.7812318 48.8557195)

359 rows × 3 columns

Figure 10-5. *A part of the data resulting from Code Block 10-5. Image by author*

This result here is a pandas dataframe. For doing spatial operations, as you know by now, it is best to convert this into a geodataframe. This can be done using the code in Code Block 10-6.

Code Block 10-6. Convert to geodataframe

```
import geopandas as gpd
gdf = gpd.GeoDataFrame(results_df, geometry='point')
gdf
```

235

Your object gdf is now georeferenced. We can move on to joining this point dataset with the store information dataset, using a spatial join. This spatial join is executed using the code in Code Block 10-7.

Code Block 10-7. Join the geodataframe to the info_polygons

```
joined_data = gpd.sjoin(gdf, info_polygons, how='left')
joined_data
```

The joined data will look as shown in Figure 10-6.

	client_id	target	point	index_right	Name	Description
0	8	Bought Yes	POINT (2.78016 48.85490)	NaN	NaN	NaN
1	8	Bought Yes	POINT (2.78018 48.85505)	NaN	NaN	NaN
2	8	Bought Yes	POINT (2.78012 48.85526)	3.0	Supermarket	
3	8	Bought Yes	POINT (2.78027 48.85533)	3.0	Supermarket	
4	8	Bought Yes	POINT (2.77994 48.85533)	3.0	Supermarket	
...	...	...	...	...	...	...
354	27	Bought No	POINT (2.78012 48.85483)	NaN	NaN	NaN
355	27	Bought No	POINT (2.78114 48.85559)	4.0	Cheap Fashion	
356	27	Bought No	POINT (2.78116 48.85549)	4.0	Cheap Fashion	
357	27	Bought No	POINT (2.78131 48.85549)	4.0	Cheap Fashion	
358	27	Bought No	POINT (2.78123 48.85572)	4.0	Cheap Fashion	

359 rows × 6 columns

Figure 10-6. *The data resulting from Code Block 10-7. Image by author*

You can see that for most points the operation has been successful. For a number of points, however, it seems that NA, or missing values, has been introduced. This is explained by the presence of points that are not overlapping with any of the store information polygons and therefore having no lookup information. It would be good to do something about this. Before deciding what to do with the NAs, let's use the code in Code Block 10-8 to count the number of each client for which there is no reference information.

Code Block 10-8. Inspect NA

```
# inspect NA
joined_data['na'] = joined_data.Name.isna()
joined_data.groupby('client_id').na.sum()
```

The result is shown in Figure 10-7.

```
client_id
8      2
9      1
10     2
11     1
12     1
13     2
14     1
15     1
16     1
17     1
18     2
19     1
20     3
21     1
22     1
23     1
24     1
25     1
26     1
27     1
Name: na, dtype: int64
```

Figure 10-7. *The result of Code Block 10-8. Image by author*

The number of nonreferenced points differs for each participant, but it never goes above three. We can therefore conclude that there is really no problem in just discarding the data that has no reference. After all, there is very little of it, and there is no added information in this data.

The code in Code Block 10-9 shows how to remove the rows of data that have missing values.

Code Block 10-9. Drop NAs

```
# drop na
joined_data = joined_data.dropna()
joined_data
```

The resulting dataframe is a bit shorter: 333 rows instead of 359, as you can see in Figure 10-8.

client_id	target		point	index_right	Name	Description	na
2	8	Bought Yes	POINT (2.78012 48.85526)	3.0	Supermarket		False
3	8	Bought Yes	POINT (2.78027 48.85533)	3.0	Supermarket		False
4	8	Bought Yes	POINT (2.77994 48.85533)	3.0	Supermarket		False
5	8	Bought Yes	POINT (2.78046 48.85529)	3.0	Supermarket		False
6	8	Bought Yes	POINT (2.78026 48.85543)	3.0	Supermarket		False
...	...	...	...	...	...	...	...
353	26	Bought No	POINT (2.78350 48.85323)	2.0	High Fashion		False
355	27	Bought No	POINT (2.78114 48.85559)	4.0	Cheap Fashion		False
356	27	Bought No	POINT (2.78116 48.85549)	4.0	Cheap Fashion		False
357	27	Bought No	POINT (2.78131 48.85549)	4.0	Cheap Fashion		False
358	27	Bought No	POINT (2.78123 48.85572)	4.0	Cheap Fashion		False

333 rows × 7 columns

Figure 10-8. *The dataset without NAs*

We now have a dataframe with explanatory information that will help us to predict coupon usage.

Reorganizing and Standardizing the Data

Now that all the needed information is there for building our classification model, we still need to work on the correct organization of this data. After all, the model interface in Python needs the data in a correct format.

In this section, we will take the spatially joined dataframe and return to a wide format, in which we again obtain a dataframe with one row per participant. Instead of having a coordinate LineString, we now want to have the participant's presence in each of the categories. We will obtain this simply by counting, for each participant, the number of points in each of the store categories. This is obtained using a groupby, which is shown in Code Block 10-10.

Code Block 10-10. The groupby to obtain location behavior

```
location_behavior = joined_data.pivot_table(index='client_id',
columns='Name', values='target',aggfunc='count').fillna(0)
location_behavior
```

The result of this groupby operation is shown in Figure 10-9.

Name client_id	Cheap Fashion	Cosmetics	Electronics	High Fashion	Restaurants	Supermarket	West Wing
8	1.0	1.0	0.0	0.0	8.0	13.0	0.0
9	1.0	1.0	0.0	0.0	0.0	13.0	1.0
10	0.0	0.0	0.0	0.0	0.0	0.0	19.0
11	2.0	0.0	1.0	0.0	18.0	1.0	0.0
12	0.0	0.0	0.0	0.0	0.0	18.0	0.0
13	1.0	0.0	1.0	0.0	6.0	6.0	6.0
14	0.0	0.0	0.0	0.0	0.0	18.0	0.0
15	0.0	0.0	0.0	0.0	0.0	0.0	12.0
16	0.0	1.0	10.0	0.0	0.0	0.0	0.0
17	1.0	1.0	11.0	0.0	17.0	0.0	0.0
18	0.0	7.0	0.0	0.0	0.0	0.0	0.0
19	7.0	9.0	0.0	0.0	0.0	0.0	0.0
20	10.0	3.0	0.0	0.0	0.0	0.0	0.0
21	0.0	2.0	0.0	20.0	1.0	0.0	0.0
22	0.0	3.0	1.0	8.0	1.0	0.0	0.0
23	0.0	7.0	1.0	12.0	1.0	0.0	0.0
24	4.0	0.0	0.0	0.0	0.0	4.0	6.0
25	0.0	1.0	1.0	5.0	6.0	0.0	0.0
26	4.0	2.0	2.0	8.0	3.0	1.0	0.0
27	4.0	0.0	0.0	0.0	0.0	0.0	0.0

Figure 10-9. *The result of Code Block 10-10. Image by author*

This grouped data already seems very usable information for understanding something about each of the clients. For example, we can see that the participant with client_id 21 has spent a huge amount of time in the High Fashion section. Another example is client_id 9, who seemed to have only gone to the supermarket.

Although this dataset is very informative, there is still one problem before moving on to the classification model. When looking at the category electronics, we can see as an example client_id 17 being the largest value inside electronics. If we look further into client_id 17, however, we see that electronics is not actually the largest category for this participant.

There is a bias in the data that is due to the fact that not all participants have the same number of points in their LineString. To solve this, we need to standardize for the number of points, which can be done using the code in Code Block 10-11.

Code Block 10-11. Standardize the data

```
# standardize
location_behavior = location_behavior.div( location_behavior.sum(axis=1),
axis=0 )
location_behavior
```

The standardized output is shown in Figure 10-10.

Name	Cheap Fashion	Cosmetics	Electronics	High Fashion	Restaurants	Supermarket	West Wing
client_id							
8	0.043478	0.043478	0.000000	0.000000	0.347826	0.565217	0.000000
9	0.062500	0.062500	0.000000	0.000000	0.000000	0.812500	0.062500
10	0.000000	0.000000	0.000000	0.000000	0.000000	0.000000	1.000000
11	0.090909	0.000000	0.045455	0.000000	0.818182	0.045455	0.000000
12	0.000000	0.000000	0.000000	0.000000	0.000000	1.000000	0.000000
13	0.050000	0.000000	0.050000	0.000000	0.300000	0.300000	0.300000
14	0.000000	0.000000	0.000000	0.000000	0.000000	1.000000	0.000000
15	0.000000	0.000000	0.000000	0.000000	0.000000	0.000000	1.000000
16	0.000000	0.090909	0.909091	0.000000	0.000000	0.000000	0.000000
17	0.033333	0.033333	0.366667	0.000000	0.566667	0.000000	0.000000
18	0.000000	1.000000	0.000000	0.000000	0.000000	0.000000	0.000000
19	0.437500	0.562500	0.000000	0.000000	0.000000	0.000000	0.000000
20	0.769231	0.230769	0.000000	0.000000	0.000000	0.000000	0.000000
21	0.000000	0.086957	0.000000	0.869565	0.043478	0.000000	0.000000
22	0.000000	0.230769	0.076923	0.615385	0.076923	0.000000	0.000000
23	0.000000	0.333333	0.047619	0.571429	0.047619	0.000000	0.000000
24	0.285714	0.000000	0.000000	0.000000	0.000000	0.285714	0.428571
25	0.000000	0.076923	0.076923	0.384615	0.461538	0.000000	0.000000
26	0.200000	0.100000	0.100000	0.400000	0.150000	0.050000	0.000000
27	1.000000	0.000000	0.000000	0.000000	0.000000	0.000000	0.000000

Figure 10-10. *The data resulting from Code Block 10-11. Image by author*

Modeling

Let's now keep the data this way for the model – for inputting the data into the model. Let's move away from the dataframe format and use the code in Code Block 10-12 to convert the data into numpy arrays.

Code Block 10-12. Convert into numpy

```
X = location_behavior.values
X
```

The X array looks something like Figure 10-11.

```
array([[0.04347826, 0.04347826, 0.        , 0.        , 0.34782609,
        0.56521739, 0.        ],
       [0.0625    , 0.0625    , 0.        , 0.        , 0.        ,
        0.8125    , 0.0625    ],
       [0.        , 0.        , 0.        , 0.        , 0.        ,
        0.        , 1.        ],
       [0.09090909, 0.        , 0.04545455, 0.        , 0.81818182,
        0.04545455, 0.        ],
       [0.        , 0.        , 0.        , 0.        , 0.        ,
        1.        , 0.        ],
       [0.05      , 0.        , 0.05      , 0.        , 0.3       ,
        0.3       , 0.3       ],
       [0.        , 0.        , 0.        , 0.        , 0.        ,
        1.        , 0.        ],
       [0.        , 0.        , 0.        , 0.        , 0.        ,
        0.        , 1.        ],
       [0.        , 0.09090909, 0.90909091, 0.        , 0.        ,
```

Figure 10-11. *The array resulting from Code Block 10-12. Image by author*

You can do the same to obtain an array for the target, also called y. This is done in Code Block 10-13.

Code Block 10-13. Get y as an array

```
y = itineraries.Name.values
y
```

The y, or target, now looks as shown in Figure 10-12.

```
array(['Bought Yes', 'Bought Yes', 'Bought Yes', 'Bought Yes',
       'Bought Yes', 'Bought Yes', 'Bought Yes', 'Bought Yes',
       'Bought Yes', 'Bought Yes', 'Bought No', 'Bought No', 'Bought No',
       'Bought No', 'Bought No', 'Bought No', 'Bought No', 'Bought No',
       'Bought No', 'Bought No'], dtype=object)
```

Figure 10-12. *The y array. Image by author*

The next step in modeling is building a train-test-split. A train-test-split in machine learning comes down to splitting your data in two, based on the rows, and using only part of your data for building the model. The part that you use for building the model is the train set, and the part that you do not use here is the test set.

The test set is important to keep apart, as machine learning models have a tendency to learn relationships that use X to predict y which are perfectly valid on the data that has been seen by the model, but that are not valid on any new data. The process is called overfitting to the train set.

When training a model on a train set and evaluating it on a dataset that was not seen by the model during the training phase, we make sure that the model evaluation is fair. We are certain that the model is not overfitted to the test data, as it was never used in the fitting (training) process.

The code in Code Block 10-14 executes a stratified train-test-split. Stratification is a form a sampling that forces the distribution of a specified variable to be the same in train and test. After all, there would be a risk that all the coupon users are in train, and then the test set would not be evaluating the performance on coupon users at all. Stratification forces the same percentage of coupon users in both train and test, which promotes fair evaluation.

Code Block 10-14. Stratified train-test-split

```
# stratified train test split
from sklearn.model_selection import train_test_split
X_train, X_test, y_train, y_test = train_test_split(X, y, test_size=0.33,
random_state=42, stratify=y)
```

After this step, you end up with four datasets: X_train and y_train are the parts of X and y that we will use for training, and X_test and y_test will be used for evaluation.

We now have all the elements to start building a model. The first model that we are going to build here is the logistic regression. As we do not have tons of data, we can exclude the use of complex models like random forests, xgboost, and the like, although they could definitely replace the logistic regression if we had more data in this use case. Thanks to the easy-to-use modeling interface of scikit-learn, it is really easy to replace one model by another, as you'll see throughout the remainder of the example.

The code in Code Block 10-15 first initiates a logistic regression and then fits the model using the training data.

Code Block 10-15. Logistic regression

```
# logistic regression
from sklearn.linear_model import LogisticRegression

my_lr = LogisticRegression()
my_lr.fit(X_train, y_train)
```

The object my_lr is now a fitted logistic regression which basically means that its coefficients have been estimated based on the training data and they have been stored inside the object. We can now use the my_lr object to make predictions on any external data that contains the same data as the one that was present in X_train.

Luckily, we have kept apart X_test so that we can easily do a model evaluation. The first step in this is to make the predictions using the code in Code Block 10-16.

Code Block 10-16. Prediction

```
preds = my_lr.predict(X_test)
preds
```

The array contains the predictions for each of the rows in X_test, as shown in Figure 10-13.

```
array(['Bought No', 'Bought Yes', 'Bought Yes', 'Bought Yes',
       'Bought Yes', 'Bought No', 'Bought No'], dtype=object)
```

Figure 10-13. *The resulting predictions. Image by author*

We do have the actual truth for these participants as well. After all, they are not really new participants, but rather a subset of participants of which we know whether they used the coupon that we chose to keep apart for evaluation. We can compare the predictions to the actual ground truth, using the code in Code Block 10-17.

Code Block 10-17. Convert the errors and ground truth to a dataframe

```
# indeed one error for the log reg
pd.DataFrame({'real': y_test, 'pred': preds})
```

The resulting comparison dataframe is shown in Figure 10-14.

	real	pred
0	Bought No	Bought No
1	Bought Yes	Bought Yes
2	Bought No	Bought Yes
3	Bought Yes	Bought Yes
4	Bought Yes	Bought Yes
5	Bought No	Bought No
6	Bought No	Bought No

Figure 10-14. *The dataframe for error analysis. Image by author*

The test set is rather small in this case, and we can manually conclude that the model is actually predicting quite well. In use cases with more data, it would be better to summarize this performance using other methods. One great way to analyze classification models is the confusion matrix. It shows in one graph all the data that is correctly predicted, but also which are wrongly predicted and in that case which errors were made how many times. The code in Code Block 10-18 shows how to create such a confusion matrix for this use case.

Code Block 10-18. Analyze the prediction errors

```
from sklearn.metrics import confusion_matrix, ConfusionMatrixDisplay

conf_mat = confusion_matrix(y_test, preds, normalize=None)
conf_mat_plot = ConfusionMatrixDisplay(conf_mat, display_labels =
set(y_test))
conf_mat_plot.plot()
```

The resulting plot is shown in Figure 10-15.

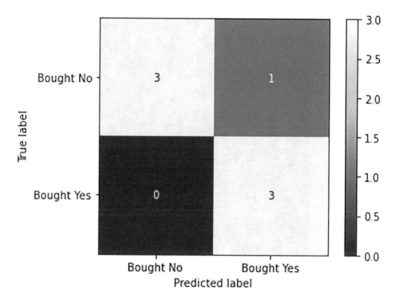

Figure 10-15. *The plot resulting from Code Block 10-18. Image by author*

In this graph, you see that most predictions were correct and only one mistake was made. This mistake was a participant that did not buy, whereas the model predicted that he was a buyer with the coupon.

Model Benchmarking

The model made one mistake, so we can conclude that it is quite a good model. However, for completeness, it would be good to try out another model. Feel free to test out any classification model from scikit-learn, but due to the relatively small amount of data, let's try out a decision tree model here. The code in Code Block 10-19 goes through the exact same steps as before but simply with a different model.

Code Block 10-19. Model benchmarking

```
from sklearn.tree import DecisionTreeClassifier
my_dt = DecisionTreeClassifier()
my_dt.fit(X_train, y_train)
preds = my_dt.predict(X_test)
pd.DataFrame({'real': y_test, 'pred': preds})
```

The resulting dataframe for evaluation is shown in Figure 10-16.

	real	pred
0	Bought No	Bought No
1	Bought Yes	Bought Yes
2	Bought No	Bought Yes
3	Bought Yes	Bought Yes
4	Bought Yes	Bought Yes
5	Bought No	Bought Yes
6	Bought No	Bought No

Figure 10-16. *The resulting dataframe from Code Block 10-19. Image by author*

Again, thanks to the small dataset size, it is easy to interpret and conclude that the model is worse, as it has made two mistakes, whereas the logistic regression only made one mistake. For coherence, let's complete this with a confusion matrix analysis as well. This is done in Code Block 10-20.

Code Block 10-20. Plot the confusion matrix

```
conf_mat = confusion_matrix(y_test, preds, normalize=None)
conf_mat_plot = ConfusionMatrixDisplay(conf_mat, display_labels =
set(y_test))
conf_mat_plot.plot()
```

The result is shown in Figure 10-17 and indeed shows two errors.

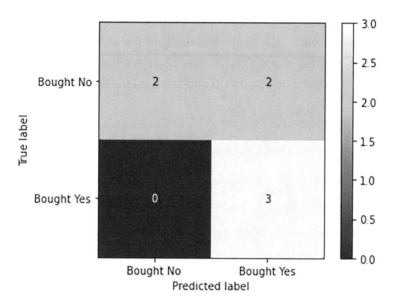

Figure 10-17. *The result from the Decision Tree model*

There are two errors, both cases of participants that did not buy in reality. It seems that people that did not buy are a little bit harder to detect than the opposite, even though more evidence would be needed to further investigate this.

Key Takeaways

1. Classification is an area in supervised machine learning that deals with models that learn how to use independent variables to predict a categorical target variable.

2. Feature engineering together with spatial operations can be used to get spatial data into a machine learning format. It is important to end up with variable definitions that will be useful for the classification task at hand.

3. Train-test-splits are necessary for model evaluation, as models tend to overfit on the training data.

4. The confusion matrix is a great tool for evaluating classification models' performances.

5. Model benchmarking is the task of using multiple different machine learning models on the same task, so that the best performing model can be found and retained for the future.

CHAPTER 11

Regression

In the previous two chapters, you have learned about the fundamentals of machine learning use cases using spatial data. You have first seen several methods of interpolation. Interpolation was presented as an introduction to machine learning, in which a theory-based interpolation function is defined to fill in unknown values of the target variable.

The next step moved from this unsupervised approach to a supervised approach, in which we build models to predict values of which we have ground truth values. By applying a train-test-split, this ground truth is then used to compute a performance metric.

The previous chapter showed how to use supervised models for classification. In classification models, unlike with interpolation, the target variable is a categorical variable. The shown example used a binary target variable, which classified people into two categories: buyers and nonbuyers.

In this chapter, you will see how to build supervised models for target variables that are numeric. This is called regression. Although regression, just like interpolation, is used to estimate a numeric target, the methods are actually generally closer to the supervised classification methods.

In regression, the use of metrics and building models with the best performance on this metric will be essential as it was in classification. The models are adapted for taking into account a numeric target variable, and the metrics need to be chosen differently to take into account the fact that targets are numeric.

The chapter will start with a general introduction of what regression models are and what we can use them for. The rest of the chapter will present an in-depth analysis of a regression model with spatial data, during which numerous theoretical concepts will be presented.

© Joos Korstanje 2022
J. Korstanje, *Machine Learning on Geographical Data Using Python*,
https://doi.org/10.1007/978-1-4842-8287-8_11

Introduction to Regression

Although the goal of this book is not to present a deep mathematical content on machine learning, let's start by exploring the general idea behind regression models anyway. Keep in mind that there are many resources that will be able to fill in this theory and that the goal of the current book is to present how regression models can be combined with spatial data analysis and modeling.

Let's start this section by considering one of the simplest cases of regression modeling: the simple linear regression. In simple linear regression, we have one numeric target variable (y variable) and one numeric predictor variable (X variable).

In this example, let's consider a dataset in which we want to predict a person's weekly weight loss based on the number of hours that a person has worked out in that same week. We expect to see a positive relationship between the two. Figure 11-1 shows the weekly weight loss plotted against the weekly workout hours.

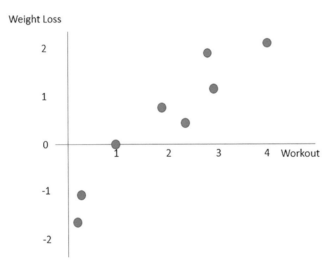

Figure 11-1. *Plot of the example data. Image by author*

This graph shows a clear positive relationship between workout and weight loss. We could find the mathematical definition of the straight line going through those points and then use this mathematical formula as a model to estimate weekly weight loss as a function of the number of hours worked out. This can be shown graphically in Figure 11-2.

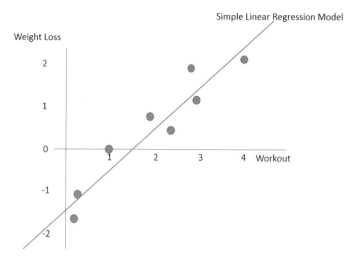

Figure 11-2. *The simple linear regression added to the graph. Image by author*

The mathematical form of this linear regression model is the following:

$$y = a * x + b$$

which would translate to the following for this example:

$$Weight\_Loss = a * Workout + b$$

Mathematical procedures to determine the best-fitting values for a and b exist and can be used to estimate this model. The exact mathematics behind this will be left for further reading as to not go out of scope for the current book. However, it is important to understand the general idea behind estimating such a model.

It is also important to consider which next steps are possible, so let's spend some time to consider those. Firstly, the current model is using only a single explanatory variable (workout), which is not really representative of how one would go about losing weight.

In reality, one could consider that the food quantity is also a very important factor in losing weight. This would need an extension of the mathematical formula to become something like the following:

$$Weight\_Loss = a * Workout + b * CaloriesEaten + c$$

In this case, the mathematics behind the linear regression would need to find best values for three coefficients: a, b, and c. This can go on by adding more variables and more coefficients.

Until here, we have firstly discussed the simple linear regression, followed by linear regression more generally. Although linear regression is often great for fitting regression models, there are many other mathematical and algorithmic functions that can be used.

Examples of other models are Decision Trees, Random Forest, and Boosting models. Deep down, each of them has their own definition of a basic model, in which the data can be used to fit the model (the alternative of estimating the coefficients in the linear model). Many reference works exist for those readers wanting to gain in-depth mathematical insights into the exact workings of those models.

For now, let's move on to a more applied vision by working through a regression use case using spatial data.

Spatial Regression Use Case

The remainder of this chapter will walk you through a number of iterations on a supervised machine learning use case. In practice, building machine learning models generally happens in multiple steps and iterations. This chapter is divided into multiple parts in order to represent this as close to reality as possible.

The goal of the model is to estimate the price that we should charge for an Airbnb location in the center of Amsterdam. Imagine that you live in Amsterdam and that you want to rent out your apartment on Airbnb for the best price, and you have collected some data to find out how to decide on the price.

The target variable of this use case is the price, and there are numerous variables that we will use as predictor variables. Firstly, there is data on the maximum number of guests that are allowed in the apartment. Secondly, data has been collected to note whether or not a breakfast is included in the price of the apartment. Let's start by importing and preparing the data.

Importing and Preparing Data

In this section, we will import and prepare the data. There are two files in the dataset:

- The geodata in a KML file

- An Excel file with price and prediction variables for each apartment

The geodataset can be imported using geopandas and Fiona, just as you have seen in earlier chapters of this book. This is done in Code Block 11-1.

Code Block 11-1. Importing the data

```
import geopandas as gpd
import fiona

gpd.io.file.fiona.drvsupport.supported_drivers['KML'] = 'rw'
geodata = gpd.read_file('chapter 11 data.kml')
geodata.head()
```

The geodata, once imported, is shown in Figure 11-3.

	Name	Description	geometry
0	Apt 1		POINT Z (4.88944 52.37603 0.00000)
1	Apt 2		POINT Z (4.88828 52.37486 0.00000)
2	Apt 3		POINT Z (4.88311 52.37163 0.00000)
3	Apt 4		POINT Z (4.88581 52.37493 0.00000)
4	Apt 5		POINT Z (4.88714 52.37163 0.00000)

Figure 11-3. *The data resulting from Code Block 11-1. Image by author*

As you can see, the dataset contains only two variables:

– The Name contains the identifier of each point.

– The Point contains the coordinates of each apartment.

The other variables are in the Excel file, which you can import using the code in Code Block 11-2.

Code Block 11-2. Importing the house data

```
import pandas as pd
apartment_data = pd.read_excel('house_data.xlsx')
apartment_data.head()
```

This second dataset is shown in Figure 11-4.

	Apt ID	Price	MaxGuests	IncludesBreakfast
0	1	120	4	0
1	2	110	2	0
2	3	115	3	0
3	4	125	4	0
4	5	135	3	0

Figure 11-4. *The house data. Image by author*

As you can see from this image, the data contains the following variables:

- Apt ID: The identifier of each apartment

- Price: The price of each apartment on Airbnb

- MaxGuest: The maximum number of guests allowed in the apartment

- IncludesBreakfast: 1 if breakfast is included and 0 otherwise

The Apt ID is not in the same format as the identifier in the geodata. It is necessary to convert the values in order to make them correspond. This will allow us to join the two datasets together in a later step. This is done using the code in Code Block 11-3.

Code Block 11-3. Convert apartment IDs

```
apartment_data['Apt ID'] = apartment_data['Apt ID'].apply(lambda x: 'Apt '
+ str(x))
apartment_data.head()
```

After this operation, the dataset now looks as shown in Figure 11-5.

	Apt ID	Price	MaxGuests	IncludesBreakfast
0	Apt 1	120	4	0
1	Apt 2	110	2	0
2	Apt 3	115	3	0
3	Apt 4	125	4	0
4	Apt 5	135	3	0

Figure 11-5. *The data resulting from Code Block 11-3. Image by author*

Now that the two datasets have an identifier that corresponds, it is time to start the merge operation. This merge will bring all columns into the same dataset, which will be easier for working with the data. This merge is done using the code in Code Block 11-4.

Code Block 11-4. Merging the two datasets

```
merged_data = geodata.merge(apartment_data, left_on='Name', right_
on='Apt ID')
merged_data.head()
```

You can see the resulting dataframe in Figure 11-6.

	Name	Description	geometry	Apt ID	Price	MaxGuests	IncludesBreakfast
0	Apt 1	POINT Z (4.88944 52.37603 0.00000)	Apt 1	120	4		0
1	Apt 2	POINT Z (4.88828 52.37486 0.00000)	Apt 2	110	2		0
2	Apt 3	POINT Z (4.88311 52.37163 0.00000)	Apt 3	115	3		0
3	Apt 4	POINT Z (4.88581 52.37493 0.00000)	Apt 4	125	4		0
4	Apt 5	POINT Z (4.88714 52.37163 0.00000)	Apt 5	135	3		0

Figure 11-6. *The merged data. Image by author*

We now have all the columns inside the same dataframe. This concludes the data preparation phase. As a last step, let's do a visualization of the apartment locations within Amsterdam, to get a better feeling for the data. This is done using the code in Code Block 11-5.

Code Block 11-5. Map of the apartments

```
import contextily as cx

# plotting all data
ax = merged_data.plot(figsize=(15,15), edgecolor='black', facecolor='none')

# adding a contextily basemap
cx.add_basemap(ax, crs=merged_data.crs)
```

The resulting map is shown in Figure 11-7.

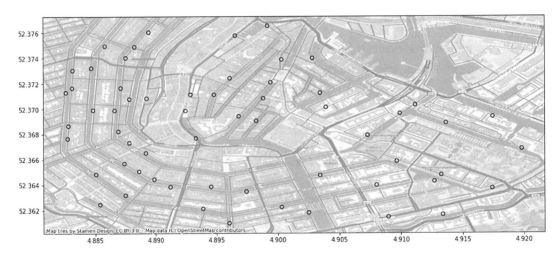

Figure 11-7. *The map resulting from Code Block 11-5. Image by author using contextily source data and image as referenced in the image*

You can see that the apartments used in this study are pretty well spread throughout the center of Amsterdam. In the next section, we will do more in-depth exploration of the dataset.

Iteration 1 of Data Exploration

In this first iteration of data exploration, let's look at how we could use the nongeographical data for a regression model. This will yield a comparable approach as the example that was given in the theoretical part earlier in this chapter.

To start the exploration, let's make a histogram of the prices of our apartments. The histogram can be created using the code in Code Block 11-6.

Code Block 11-6. Creating a histogram

```
import matplotlib.pyplot as plt
plt.hist(merged_data['Price'])
```

The result of this histogram is shown in Figure 11-8.

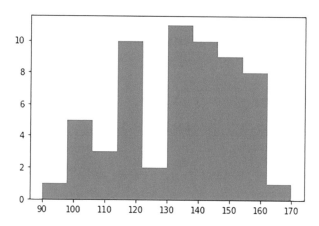

Figure 11-8. *Histogram of the prices. Image by author*

This histogram shows us that the prices are all between 90 and 170, with the majority being at 130. The data does not seem to follow a perfectly normal distribution, although we do see more data points being closer to the center than further away.

If we would need to give a very quick-and-dirty estimation of the most appropriate estimation for the price of our Airbnb, we could simply use the average price of Airbnbs in the center of Amsterdam. The code in Code Block 11-7 computes this mean.

Code Block 11-7. Compute the mean

```
# if we had no info to segment at all, our best guest would be to predict
the mean
merged_data['Price'].mean()
```

The result is 133.75, which tells us that setting this price would probably be a more or less usable estimate if we had nothing more precise. Of course, as prices range from 90 to 170, we could be either:

  – Lose money due to underpricing: If our Airbnb is actually worth 170 and we choose to price it at 133.75, we would be losing the difference (170 – 133.75) each night.

– Lose money due to overpricing: If our Airbnb is actually worth 90 and we choose to price it at 133.75, we will probably have a very hard time finding guests, and our booking number will be very low.

Clearly, it would be very valuable to have a better understanding of the factors influencing Airbnb price so that we can find the best price for our apartment.

As a next step, let's find out how the number of guests can influence Airbnb prices. The code in Code Block 11-8 creates a scatter plot of Price against MaxGuests to visually inspect relationships between those variables.

Code Block 11-8. Create a scatter plot

```
# however we may use additional information to make this estimate
more fitting
plt.scatter(merged_data['MaxGuests'], merged_data['Price'])
```

Although the trend is less clear than the one observed in the theoretical example in the beginning of this chapter, we can clearly see that higher values on the x axis (MaxGuests) generally have higher values on the y axis (Price). Figure 11-9 shows this.

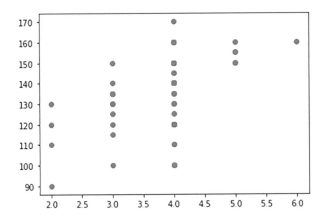

Figure 11-9. *The scatter plot of Price against MaxGuests. Image by author*

The quality of a linear relationship can also be measured using a more quantitative approach. The Pearson correlation coefficient is a sort of score between –1 and 1 that gives this indication. A value of 0 means no correlation, a value close to –1 means a negative correlation between the two, and a value close to 1 means a positive correlation between the variables.

The correlation coefficient can be computed using the code in Code Block 11-9.

Code Block 11-9. Compute the correlation coefficient

```
import numpy as np
np.corrcoef(merged_data['MaxGuests'], merged_data['Price'])
```

This will give you the correlation matrix as shown in Figure 11-10.

```
array([[1.        , 0.45366546],
       [0.45366546, 1.        ]])
```

Figure 11-10. *The correlation coefficient. Image by author*

The resulting correlation coefficient between MaxGuests and Price is 0.453. This is a fairly strong positive correlation, indicating that the number of guests has a strong positive impact on the price that we can ask for an Airbnb. In short, Airbnbs for more people should ask a higher price, whereas Airbnbs for small number or guests should price lower.

As a next step, let's see whether we can also use the variable IncludesBreakfast for setting the price of our Airbnb. As the breakfast variable is categorical (yes or no), it is better to use a different technique for investigating this relationship. The code in Code Block 11-10 creates a boxplot to answer this question.

Code Block 11-10. Create a boxplot

```
import seaborn as sns
sns.boxplot(x='IncludesBreakfast',y='Price',data=merged_data)
```

The resulting boxplot is shown in Figure 11-11.

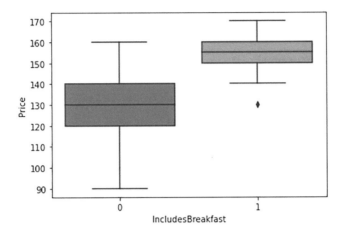

Figure 11-11. *The resulting boxplot. Image by author*

This boxplot shows us that Airbnbs that propose a breakfast are generally able to ask a higher price than Airbnbs that do not propose one. Depending on whether you propose a breakfast, you should price your apartment accordingly.

Iteration 1 of the Model

The insights from the previous section will already be very useful elements for an apartment owner. However, they remain a more general advice rather than an exact numerical estimation of the price that we'd need to set. In this section, we'll build a first version of a machine learning model to set this price exactly based on the variables breakfast and MaxGuest. As a first set, let's set X and y as variables for modeling. This is done in Code Block 11-11.

Code Block 11-11. Creating X and y objects

```
X = merged_data[['IncludesBreakfast', 'MaxGuests']]
y = merged_data['Price']
```

We will use a linear model for this phase of modeling. The scikit-learn implementation of the linear model can be estimated using the code in Code Block 11-12.

Code Block 11-12. Linear regression

```
# first version lets just do a quick and dirty non geo model
from sklearn.linear_model import LinearRegression
lin_reg_1 = LinearRegression()
lin_reg_1.fit(X, y)
```

Now that the model has been fitted, we have the mathematical definition (with the estimated coefficients) inside our linear regression object.

Interpretation of Iteration 1 Model

To interpret what this model has learned, we can inspect the coefficients. The code in Code Block 11-13 shows how to see the coefficients that the model has estimated.

Code Block 11-13. Print the interpretation of the linear model

```
print('When no breakfast and 0 Max Guests then price is estimated at: ',
lin_reg_1.intercept_)

print('Adding breakfast adds to the price: ', lin_reg_1.coef_[0])

print('Each additional Max Guests adds to the price: ', lin_reg_1.coef_[1])
```

This code results in the following output:

- When no breakfast and 0 Max Guests then price is estimated at:
  103.1365444728616

- Adding breakfast adds to the price: 16.615515771525995

- Each additional Max Guests adds to the price: 7.254546746234728

These coefficients give a precise mathematical estimation of the insights that we obtained in the previous section. Although these numbers seem very precise, this does not mean that they are actually correct. It is important to define a metric and measure whether this model is any good.

In the following, we will compute the R2 score, which is a regression metrics that generally falls between 0 (model has no value) and 1 (perfect model). Values can sometimes fall below 0, indicating that the model has no value at all.

The R2 score can be computed on all data, but this is not the preferred way for estimating model performance. Machine learning models tend to learn very well on the data that was seen during training (fitting), without necessarily generalizing very well. A solution for this is to split the data in a training set (observations that are used for the training/estimation) and a test set that is used for model evaluation.

The code in Code Block 11-14 splits that initial data into a training and a test set.

Code Block 11-14. Split the data in train and test

```
# Evaluate this model a bit better with train test
from sklearn.model_selection import train_test_split
X_train, X_test, y_train, y_test = train_test_split(X, y, test_size=0.33,
random_state=42)
```

Let's now fit the model again, but this time only on the training data. This is done in Code Block 11-15.

Code Block 11-15. Fit the model on the train data

```
lin_reg_2 = LinearRegression()
lin_reg_2.fit(X_train, y_train)
```

To estimate the performance, we use the estimated model (in this case, the coefficients and the linear model formula) to predict estimate prices on the test data. This is done in Code Block 11-16.

Code Block 11-16. Predict on the test set

```
pred_reg_2 = lin_reg_2.predict(X_test)
```

We can use these predicted values together with the real, known prices of the test set to compute the R2 scores. This is done in Code Block 11-17.

Code Block 11-17. Compute the R2 score

```
from sklearn.metrics import r2_score
r2_score(y_test, pred_reg_2)
```

The resulting R2 score is 0.1007. Although not a great result, the score shows that the model has some predictive value and would be a better segmentation than using the mean for pricing.

Iteration 2 of Data Exploration

In this second iteration, we will add the geographical data into the exploration and model. After all, location is important when booking an Airbnb, and this will probably be translated in price. Let's start by looking into more detail into these features of the dataset. The code in Code Block 11-18 creates variables specifically for latitude and longitude by extracting this from the geometry column.

Code Block 11-18. Add the geographic data

```
# add the geo data and see whether it improves thing
merged_data['long'] = merged_data['geometry'].apply(lambda x: x.x)
merged_data['lat'] = merged_data['geometry'].apply(lambda x: x.y)
merged_data.head()
```

The resulting dataframe is shown in Figure 11-12.

	Name	Description	geometry	Apt ID	Price	MaxGuests	IncludesBreakfast	long	lat
0	Apt 1		POINT Z (4.88944 52.37603 0.00000)	Apt 1	120	4	0	4.889439	52.376029
1	Apt 2		POINT Z (4.88828 52.37486 0.00000)	Apt 2	110	2	0	4.888281	52.374863
2	Apt 3		POINT Z (4.88311 52.37163 0.00000)	Apt 3	115	3	0	4.883109	52.371628
3	Apt 4		POINT Z (4.88581 52.37493 0.00000)	Apt 4	125	4	0	4.885813	52.374929
4	Apt 5		POINT Z (4.88714 52.37163 0.00000)	Apt 5	135	3	0	4.887143	52.371628

Figure 11-12. *The dataset resulting from Code Block 11-18. Image by author*

Let's see how latitude and longitude are related to the price by making scatter plots of price vs. latitude and price vs. longitude. The first scatter plot is created in Code Block 11-19.

Code Block 11-19. Create the scatter plot

```
plt.scatter(merged_data['lat'], merged_data['Price'])
```

The resulting scatter plot is shown in Figure 11-13.

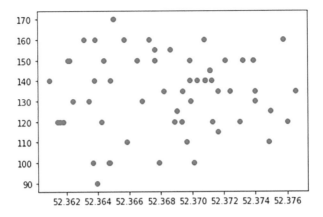

Figure 11-13. *The scatter plot resulting from Code Block 11-19. Image by author*

There does not seem to be too much of a trend in this scatter plot. It seems that prices are ranging between 90 and 170, and that is not different for any other latitude. Let's use the code in Code Block 11-20 to check whether this is true for longitude as well.

Code Block 11-20. Create a scatter plot with longitude

```
plt.scatter(merged_data['long'], merged_data['Price'])
```

The resulting scatter plot is shown in Figure 11-14.

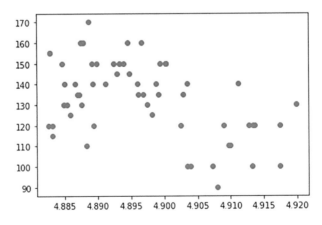

Figure 11-14. *The scatter plot of price vs. longitude. Image by author*

Interestingly, the relationship between price and longitude is much stronger. It seems that higher longitudes generally have a lower price than lower longitudes.

These two scatter plots do not really capture location. After all, we could easily imagine that relationships with latitude and longitude are not necessarily linear. It would be weird to expect that the more you go to the east, the lower your price, is a rule that always holds. It is more likely that there are specific high-value and low-value areas within the overall area. In the following code, we create a visualization that plots prices based on latitude and longitude at the same time.

The first step in creating this visualization is to convert price into a variable that can be used to set point size for the apartments. This requires our data to be scaled into a range that is more appropriate for plotting, for example, setting the cheapest apartments to a point size of 16 and the most expensive ones to a point size of 512. This is done in Code Block 11-21.

Code Block 11-21. Apply a MinMaxScaler

```
from sklearn.preprocessing import MinMaxScaler
scaler = MinMaxScaler(feature_range=(16, 512))
merged_data[['MarkerSize']] = scaler.fit_transform(merged_data[['Price']])
```

We can now use this size setting when creating the scatter plot. This is done in Code Block 11-22.

Code Block 11-22. Create a map with size of marker

```
plt.scatter(merged_data['long'], merged_data['lat'], s=merged_data['MarkerSize'], c='none', edgecolors='black')
```

The resulting visualization is shown in Figure 11-15.

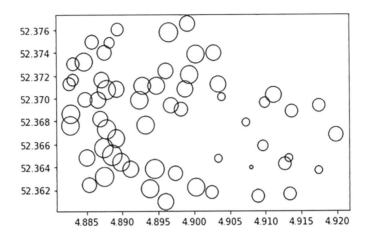

Figure 11-15. *The resulting scatter plot*

This graph shows that there are no linear relationships, but that we could expect some areas to be learned that have generally high or generally low prices. This would mean that we may need to change for a nonlinear model to fit this reality better.

Iteration 2 of the Model

Although the previous exploration just told us that the go-to model is probably nonlinear, let's do a final pass with a linear model. This will give us a good benchmark R2 score for evaluating the remaining trials. This is done in Code Block 11-23.

Code Block 11-23. Model iteration 2

```
# add features
X2 = merged_data[['IncludesBreakfast', 'MaxGuests', 'lat', 'long']]
y = merged_data['Price']

# train test split
X2_train, X2_test, y_train, y_test = train_test_split(X2, y, test_
size=0.33, random_state=42)

# build the model
lin_reg_3 = LinearRegression()
lin_reg_3.fit(X2_train, y_train)
```

```
# evaluate the model
pred_reg_3 = lin_reg_3.predict(X2_test)
print(r2_score(y_test, pred_reg_3))
```

The R2 score that is obtained by this model is 0.498. This is actually quite an improvement compared to the first iteration (which was at R2 of 0.10). This gives confidence in moving on to the tests for nonlinear model in the next section.

Iteration 3 of the Model

As already described in the introduction, there are many alternative models that we could use in the next iterations. In order to keep a model that is easy to interpret, we will use a Decision Tree in this step. Feel free to try out more complex models from the scikit-learn library, which will be just as easy to plug into this code as the following example. The code for the DecisionTreeRegressor is shown in Code Block 11-24.

Code Block 11-24. Model iteration 3

```
from sklearn.tree import DecisionTreeRegressor

# build the model
dt_reg_4 = DecisionTreeRegressor()
dt_reg_4.fit(X2_train, y_train)

# evaluate the model
pred_reg_4 = dt_reg_4.predict(X2_test)
print(r2_score(y_test, pred_reg_4))
```

The score that this model obtains is –0.04. Unexpectedly, we have a much worse result than in the previous step. Be careful here, as the DecisionTree results will be different for each execution due to randomness in the model building phase. You will probably have a different result than the one presented here, but if you try out different runs, you will see that the average performance is worse than the previous iteration.

The DecisionTreeRegressor, just like many other models, can be tuned using a large number of hyperparameters. In this iteration, no hyperparameters were specified, which means that only default values were used.

As we have a strong intuition that nonlinear models should be able to obtain better results than a linear model, let's play around with hyperparameters in the next iteration.

Iteration 4 of the Model

Max_depth is an important hyperparameter for the DecisionTreeRegressor. If you do not specify a max depth, the estimated decision tree can become very complex. Since the score in the previous iteration was so much worse than the linear model, we could expect that our decision tree became too complex. The loop in Code Block 11-25 tests out different values for max_depth, which will allow us to see whether this would give a better performing model.

Code Block 11-25. Tuning the model with max_depth

```
# tune this model a little bit
for max_depth in range(1,11):

    # build the model
    dt_reg_5 = DecisionTreeRegressor(max_depth=max_depth)
    dt_reg_5.fit(X2_train, y_train)

    # evaluate the model
    pred_reg_5 = dt_reg_5.predict(X2_test)
    print(max_depth, r2_score(y_test, pred_reg_5))
```

The resulting output is shown in Figure 11-16.

```
1 0.4132332385169546
2 0.3524351192047155
3 0.5448849007360488
4 0.4387428700696431
5 0.15716547901821065
6 0.08656285739421166
7 -0.04829770387965171
8 0.09738717339667458
9 0.18289786223277915
10 -0.05463182897862229
```

Figure 11-16. *The result of the model tuning loop. Image by author*

In this output, you can see that the max_depth of 3 has resulted in an R2 score of 0.54, much better than the result of –0.04. Tuning on max_depth has clearly had an important impact on the model's performance. Many other trials and iterations would be possible, but that is left as an exercise. For now, the DecisionTreeRegressor with max_depth = 3 is retained as the final regression model.

Interpretation of Iteration 4 Model

As a final step for this use case, let's have a closer look at what is learned by this nonlinear regression model. The great advantage of the Decision Tree is that we can export a plot of the Decision Tree that shows us exactly how the model has learned its trends. The code in Code Block 11-26 shows how to generate the tree plot.

Code Block 11-26. Generate the tree plot

```
from sklearn import tree

# build the model
dt_reg_5 = DecisionTreeRegressor(max_depth=3)
dt_reg_5.fit(X2_train, y_train)

plt.figure(figsize=(15,15))
tree.plot_tree(dt_reg_5, feature_names=X2_train.columns)
plt.show()
```

The result is shown in Figure 11-17.

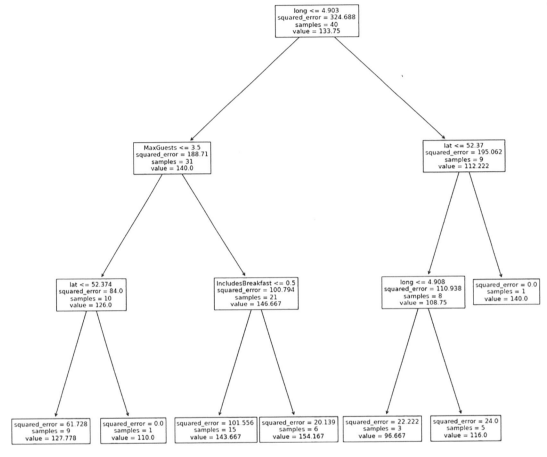

Figure 11-17. *The tree plot resulting from Code Block 11-26. Image by author*

We can clearly see which nodes have learned which trends. We can see that latitude and longitude are used multiple times by the model, which allows the model to split out specific areas on the map that are to be prices worse or better.

As this is the final model for the current use case, and we know that the R2 score tells us that the model is a much better estimation than using just an average price, we can be confident that pricing our Airbnb using the decision tree model will result in a more appropriate price for our apartment.

The goal of the use case has therefore been reached: we have created a regression model to use both spatial data and apartment data to make the best possible price estimation for an Airbnb in Amsterdam.

Key Takeaways

1. Regression is an area in supervised machine learning that deals with models that learn how to use independent variables to predict a numeric target variable.

2. Feature engineering, spatial data, and other data can be used to feed this regression model.

3. The R2 score is a metric that can be used for evaluation regression models.

4. Linear regression is one of the most common regression models, but many alternative models, including Decision Tree, Random Forest, or Boosting, can be used to challenge its performances in a model benchmark.

CHAPTER 12

Clustering

In this fourth and last chapter on machine learning, we will cover clustering. To get this technique in perspective, let's do a small recap of what we have gone through in terms of machine learning until now.

The machine learning topics started after the introduction of interpolation. In interpolation, we tried to estimate a target variable for locations at which the value of this target variable is unknown. Interpolation uses a mathematical formula to decide on the best possible theoretical way to interpolate these values.

After interpolation, we covered classification and regression, which are the two main categories in supervised modeling. In supervised modeling, we build a model that uses X variables to predict a target (y) variable. The great thing about supervised models is that we have a large number of performance metrics available that can help us in tuning and improving the model.

Introduction to Unsupervised Modeling

In this chapter, we will go deeper into unsupervised models. They are the opposite of supervised models in the sense that there is no notion of target variable in unsupervised models. There are two main types inside unsupervised models:

1. Feature reduction

2. Clustering

In feature reduction, the goal is to take a dataset with a large number of variables and then redefine these variables in a more efficient variable definition. Especially when many of the variables are strongly correlated, you can reduce the number of variables in the dataset in such a way that the new variables are not correlated.

J. Korstanje, *Machine Learning on Geographical Data Using Python*,
https://doi.org/10.1007/978-1-4842-8287-8_12

Feature reduction will be a great first step for machine learning data preprocessing and can also be used for data analysis. Examples of methods are PCA, Factor Analysis, and more. Feature reduction is not much different on geospatial data than on regular data, which is why we will not dedicate more space for this technique.

A second family of models within unsupervised models is clustering. Clustering is very different from feature reduction, except from the fact that the notion of target variable is absent in both types of models. Clustering on spatial data is quite different from clustering on regular data, which is why this chapter will present clustering on geodata in depth.

Introduction to Clustering

In clustering, the goal is to identify clusters, or groups, of observations based on some measure of similarity or distance. As mentioned before, there is no target variable here: we simply use all of the available variables about each observation to create groups of similar observations.

Let's consider a simple and often used example. In the graph in Figure 12-1, you'll see a number of people (each person is an observation) of which we have collected the spending on two product groups at a supermarket: snacks and fast food is the first category and healthy products is the second.

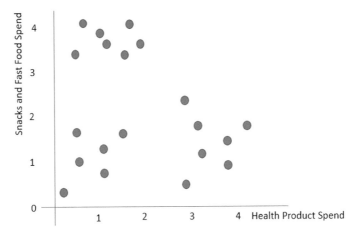

Figure 12-1. *The graph showing the example. Image by author*

As this data has only two variables, it is relatively easy to identify three groups of clients in this database. A subjective proposal for boundaries is presented in the graph in Figure 12-2.

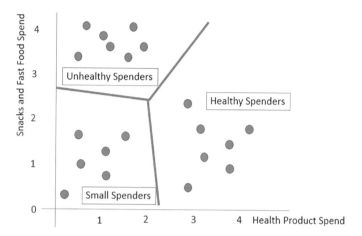

Figure 12-2. *The graph showing a clustering example. Image by author*

In this graph, you see that the clients have been divided in three groups:

1. Unhealthy spenders: A cluster of clients who spend a lot in the category snacks and fast food, but not much in the category healthy

2. Healthy spenders: A cluster of clients who spend a lot on healthy products but not much on snacks and fast food

3. Small spenders: People who do not spend a lot at all

An example of a way in which a supermarket could use such a clustering is sending personalized advertisements or discount coupons to those clients of which they know they'll be interested.

Different Clustering Models

Now, instead of doing this manual split on the graph, we will need a more precise mathematical machine learning model to define such splits. Luckily, a large number of such models exist. Examples are

  – The k-means algorithm

  – Hierarchical clustering

- DBSCAN

- OPTICS

- Gaussian mixture

- And many more. The following source contains a rich amount of information on different clustering methods: `https://scikit-learn.org/stable/modules/clustering.html`.

Many of those models, however, are unfortunately not usable for spatial data. The problem with most models is that they compute Euclidean distances between two data points or other distance and similarity metrics.

In spatial data, as covered extensively in this book, we work with latitude and longitude coordinates, and there are specific things to take into account when computing distances from one coordinate to the other. Although we could use basic clustering models as a proxy, this would be wrong, and we'd need to hope that the impact is not too much. It will be better to choose clustering methods that are specifically designed for spatial data and that can take into account correct measures of distance.

The main thing that you need to consider very strongly in clustering on spatial data is the distance metric that you are going to use. There is no one-size-fits-all method here, but we'll discover such approaches and considerations throughout the spatial clustering use case that is presented in this chapter.

Spatial Clustering Use Case

In the remainder of this chapter, we will go through a use case for clustering on spatial data. The use case is the identification on points of interest based solely on GPS tracks of people.

With those tracking points, we will try to identify some key locations during their trip, based on the idea that people have spent a little time at that location and therefore will have multiple points in a cluster of interest. On the other hand, the points will be more spread out and less clustered when people are on transportation between points.

We can use clustering for this, since we can expect clustering to find the clusters representing the points of interest. After this, we will use the center points of the identified clusters as an estimate of the real point of interest.

The steps that we will take from here are

- Importing and inspecting the data

- Building a cluster model for one user

- Tuning the clustering model for this user

- Applying the tuned model to the other users

Let's start by importing the data in the next section.

Importing and Inspecting the Data

Let's get started by importing and inspecting the data. The data is provided as a KML file. You can use the code in Code Block 12-1 to import the data into Python.

Code Block 12-1. Importing the data

```
import geopandas as gpd
import fiona

gpd.io.file.fiona.drvsupport.supported_drivers['KML'] = 'rw'
geodata = gpd.read_file('chapter_12_data.kml')
geodata.head()
```

The geodataframe looks as shown in Figure 12-3.

	Name	Description	geometry
0	Person 1		LINESTRING Z (4.31777 50.86752 0.00000, 4.3180...
1	Person 2		LINESTRING Z (4.35425 50.86111 0.00000, 4.3542...
2	Person 3		LINESTRING Z (4.29375 50.83459 0.00000, 4.2947...

Figure 12-3. *The data. Image by author*

The data are stored as one LineString for each person. There are no additional variables available. Let's now make a simple plot to have a better idea of the type of trajectories that we are working with. This can be done using the code in Code Block 12-2.

Code Block 12-2. Plotting the data

```
geodata.plot()
```

The resulting plot is shown in Figure 12-4.

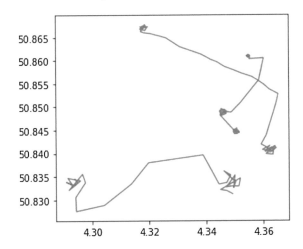

Figure 12-4. *The plotted trajectories. Image by author*

To add a bit of context to these trajectories, we can add a background map to this graph using the code in Code Block 12-3.

Code Block 12-3. Plotting with a background map

```
import contextily as cx
ax = geodata.plot(figsize=(15,15), markersize=64)
cx.add_basemap(ax, crs = geodata.crs)
```

The resulting map is shown in Figure 12-5.

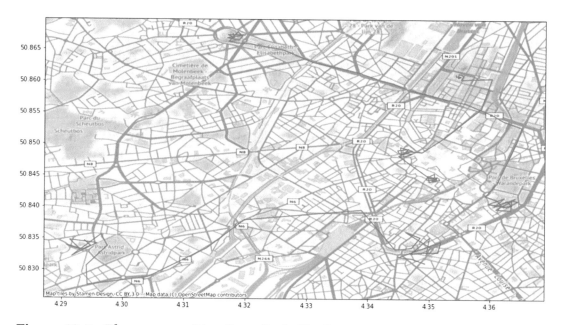

Figure 12-5. *The map resulting from Code Block 12-3. Image by author using contextily source data and image as referenced in the image*

The three trajectories are based in the city of Brussels. For each of the three trajectories, you can visually identify a similar pattern: there are clustered parts where there are multiple points in the same neighborhood, indicating points of interest. Then there are also some parts where there is a real line-like pattern which indicates movements from one point of interest to another.

Cluster Model for One Person

Let's now move on to the machine learning part. In this section, we start by extracting only one trajectory, so one person. We will apply a clustering to this trajectory to see if it can identify clusters in this trajectory.

We start by extracting the data for the first person using the code in Code Block 12-4.

Code Block 12-4. Extract data of Person 1

```
#Let's start with finding POI of one person
one_person = geodata[geodata['Name'] =='Person 1']
one_person
```

The data for Person 1 looks as shown in Figure 12-6.

	Name	Description	geometry
0	Person 1		LINESTRING Z (4.31777 50.86752 0.00000, 4.3180...

Figure 12-6. *The result from Code Block 12-4. Image by author*

Let's plot the trajectory of this person in order to have a more detailed vision of the behavior of this person. This map can be made using the code in Code Block 12-5.

Code Block 12-5. Creating a map of the trajectory of Person 1

```
ax = one_person.plot(figsize=(15,15), markersize=64)
cx.add_basemap(ax, crs = one_person.crs)
```

The resulting map is shown in Figure 12-7.

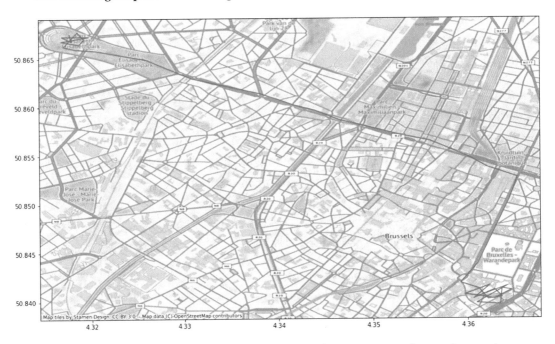

Figure 12-7. *The map resulting from Code Block 12-5. Image by author using contextily source data and image as referenced in the image*

You can see from this visualization that the person has been at two locations for a longer period: one location on the top left of the map and a second point of interest on the bottom right. We want to reach a clustering model that is indeed capable of capturing these two locations.

To start building a clustering model for Person 1, we need to convert the LineString into points. After all, we are going to cluster individual points to identify clusters of points. This is done using the code in Code Block 12-6.

Code Block 12-6. Convert the LineString into points

```
import pandas as pd
one_person_points_df = pd.DataFrame(
    [x.strip('(').strip(')').strip('0').strip(' ').split(' ')
    for x in str(one_person.loc[0, 'geometry'])[13:].split(',')],
    columns=['long','lat']
)
one_person_points_df = one_person_points_df.astype(float)
one_person_points_df.head()
```

The data format that results from this code is shown in Figure 12-8.

	long	lat
0	4.317766	50.867523
1	4.318023	50.866765
2	4.316993	50.867253
3	4.319826	50.867253
4	4.316908	50.866765

Figure 12-8. *The new data format of latitude and longitude as separate columns. Image by author*

Now that we have the right data format, it is time to apply a clustering method. As our data is in latitude and longitude, the distance between two points should be defined using haversine distance. We choose to use the OPTICS clustering method, as it applies well to spatial data. Its behavior is the following:

- OPTICS decides itself on the number of clusters that it wants to use. This is opposed to a number of models in which the user has to decide on the number of clusters.

- OPTICS can be tuned to influence the number of clusters that the model chooses. This is important, as the default settings may not result in the exact number of clusters that we want to obtain.

- OPTICS is able to discard points: when points are far away from all identified clusters, they can be coded as –1, meaning an outlier data point. This will be important in the case of spatial clustering, as there will be many data points that are on a transportation part of the trajectory that will be quite far away from the cluster centers. This option is not available in all clustering methods, but it is there in OPTICS and some other methods like DBSCAN.

Let's start with an OPTICS clustering that uses the default settings. This is done in the code in Code Block 12-7.

Code Block 12-7. Apply the OPTICS clustering

```
from sklearn.cluster import OPTICS
import numpy as np

clustering = OPTICS(metric='haversine')

one_person_points_df.loc[:,'cluster'] = clustering.fit_predict(np.
radians(one_person_points_df[['lat', 'long']]))
```

The previous code has created a column called cluster in the dataset, which now contains the cluster that the model has found for each row, each data point. The code in Code Block 12-8 shows how to have an idea of how the clusters are distributed.

Code Block 12-8. Show the value counts

```
one_person_points_df['cluster'].value_counts()
```

The result is shown in Figure 12-9.

```
       0      13
       1      12
      -1       9
       2       7
Name: cluster, dtype: int64
```

Figure 12-9. *The result of the OPTICS clustering. Image by author*

Now, as said before, the cluster –1 identified outliers. Let's delete them from the data with the code in Code Block 12-9.

Code Block 12-9. Remove the outlier cluster

```
# remove all the observations with cluster -1 (outliers)
one_person_points_df = one_person_points_df[one_person_points_
df['cluster'] != -1]
```

We can now compute the central points of each cluster by computing the median point with a groupby operation. This is done in Code Block 12-10.

Code Block 12-10. Compute medians of clusters

```
medians_of_POI = one_person_points_df.groupby(['cluster'])[['lat',
'long']].median().reset_index(drop=False)
medians_of_POI
```

Figure 12-10 shows the median coordinate for clusters 0, 1, and 2.

	cluster	lat	long
0	0	50.866765	4.318624
1	1	50.857501	4.351240
2	2	50.840377	4.362827

Figure 12-10. *The resulting clusters. Image by author*

Let's plot those central coordinates on a map using the code in Code Block 12-11.

Code Block 12-11. Plotting the central coordinates of Person 1

```
from shapely.geometry.point import Point
medians_of_POI_gdf = gpd.GeoDataFrame(medians_of_POI, geometry=[Point(x)
for x in zip( list(medians_of_POI['long']), list(medians_of_POI['lat']))])
medians_of_POI_gdf.plot()
```

The basic plot with the three central points is shown in Figure 12-11.

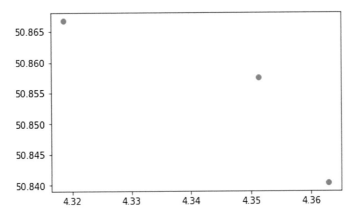

Figure 12-11. *The plot with the three central points of Person 1. Image by author*

Let's use the code in Code Block 12-12 to add more context to this map.

Code Block 12-12. Plot a basemap behind the central points

```
ax = one_person.plot(figsize=(15,15))
medians_of_POI_gdf.plot(ax=ax,markersize=128)
cx.add_basemap(ax, crs = one_person.crs)
```

The result is shown in Figure 12-12.

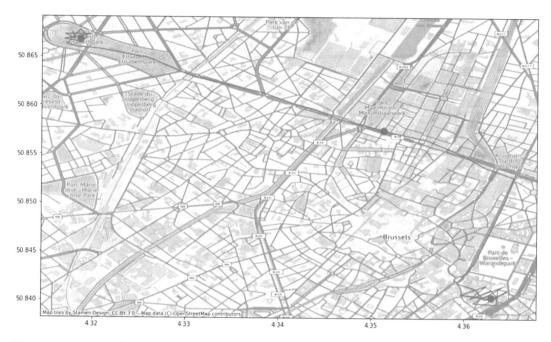

Figure 12-12. *Plotting the central points to a background map. Image by author using contextily source data and image as references in the image*

This map shows that the clustering was not totally successful. The cluster centroid on the top left did correctly identify a point of interest, and the one bottom right as well. However, there is one additional centroid in the middle that should not have been identified as a point of interest. In the next section, we will tune the model to improve this result.

Tuning the Clustering Model

Tuning models is much more complicated in unsupervised models than in supervised models. After all, we do not have an annotated dataset: the "real" points of interest are not known. The best thing is to do some manual checks and keep in mind what your objective really is. In the current case, we want to find centroids, and only centroids of points of interest.

In the following code, a different setting has been set in the OPTICS model:

– Min_samples is set to 10.

The number of samples in a neighborhood for the point to be considered a core point.

– Max_eps is set to 2.

This hyperparameter indicates the maximum distance for a point to a cluster centroid that is allowed to still be considered part of the cluster.

– Min_cluster_size is set to 8.

The minimum number of data points that have to be in a cluster if the cluster is to be kept.

– Xi is set to 0.05.

Helps in tuning the number of clusters.

These hyperparameter values have been obtained by trying out different settings and then looking whether the identified centroids coincided with the points of interest on the map. The code is shown in Code Block 12-13.

Code Block 12-13. Applying the OPTICS with different settings

```
# try different settings
one_person_points_df = pd.DataFrame(
    [x.strip('(').strip(')').strip('0').strip(' ').split(' ')
    for x in str(one_person.loc[0, 'geometry'])[13:].split(',')],
    columns=['long','lat']
)
one_person_points_df = one_person_points_df.astype(float)

clustering = OPTICS(
     min_samples = 10,
     max_eps=2.,
     min_cluster_size=8,
     xi = 0.05,
     metric='haversine')
```

```
one_person_points_df.loc[:,'cluster'] = clustering.fit_predict(
        np.radians(one_person_points_df[['lat', 'long']]))
one_person_points_df =
        one_person_points_df[one_person_points_df['cluster'] != -1]
medians_of_POI = one_person_points_df.groupby(['cluster'])[['lat',
'long']].median().reset_index(drop=False)

print(medians_of_POI)

medians_of_POI_gdf = gpd.GeoDataFrame(medians_of_POI,
        geometry=
                [Point(x) for x in
                    zip(
                            list(medians_of_POI['long']),
                            list(medians_of_POI['lat'])
                    )
                ])
ax = one_person.plot(figsize=(15,15))
medians_of_POI_gdf.plot(ax=ax,markersize=128)
cx.add_basemap(ax, crs = one_person.crs)
```

The result of this new clustering is shown in Figure 12-13.

```
   cluster        lat       long
0        0  50.866223   4.319311
1        1  50.841082   4.362312
```

Figure 12-13. *The map resulting from Code Block 12-13. Image by author using contextily data and image as referenced in the map*

As you can see, the model has correctly identified the two points (top left and bottom right) and no other points. The model is therefore successful at least for this person. In the next section, we will apply this to the other data as well and see whether the new cluster settings give correct results for them as well.

Applying the Model to All Data

Let's now loop through the three people in the dataset and apply the same clustering method for each of them. For each person, the cluster centroids will be printed and plotted on a map against the original trajectory. This will allow us to check whether the result is correct. The code to do all this is shown in Code Block 12-14.

Code Block 12-14. Apply the model to all data

```
import matplotlib.pyplot as plt

for i,row in geodata.iterrows():
  print(row)
  one_person_points_df = pd.DataFrame(
    [x.strip('(').strip(')').strip('0').strip(' ').split(' ')
     for x in str(row['geometry'])[13:].split(',')],
    columns=['long','lat']
  )
  one_person_points_df = one_person_points_df.astype(float)

  clustering = OPTICS(
                min_samples = 10,
                max_eps=2.,
                min_cluster_size=8,
                xi = 0.05,
                metric='haversine')

  one_person_points_df.loc[:,'cluster'] = clustering.fit_predict(
      np.radians(one_person_points_df[['lat', 'long']]))

  one_person_points_df =
      one_person_points_df[one_person_points_df['cluster'] != -1]

  medians_of_POI =
      one_person_points_df.groupby(['cluster'])[['lat', 'long']].median().
      reset_index(drop=False)

  print(medians_of_POI)

  medians_of_POI_gdf = gpd.GeoDataFrame(medians_of_POI,
          geometry=
              [Point(x) for x in
                  zip(
                      list(medians_of_POI['long']),
                      list(medians_of_POI['lat'])
```

```
                              )
                      ])
  ax = gpd.GeoDataFrame([row],
       geometry=[row['geometry']]).plot(figsize=(15,15))

  medians_of_POI_gdf.plot(ax=ax,markersize=128)
  plt.show()
```

The resulting output and graphs will be shown hereafter in Figures 12-14, 12-15, and 12-16.

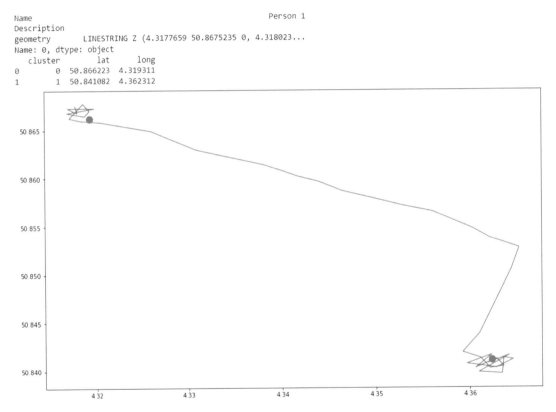

Figure 12-14. *Central points on trajectory of Person 1. Image by author*

This first map shows the result that we have already used before. Indeed, for Person 1, the OPTICS model has correctly identified the two points of interest. Figure 12-15 shows the results for Person 2.

```
Name                                           Person 2
Description
geometry        LINESTRING Z (4.3542496 50.8611072 0, 4.354249...
Name: 1, dtype: object
   cluster        lat      long
0        0  50.860809  4.355065
1        1  50.848713  4.345924
2        2  50.844689  4.350430
```

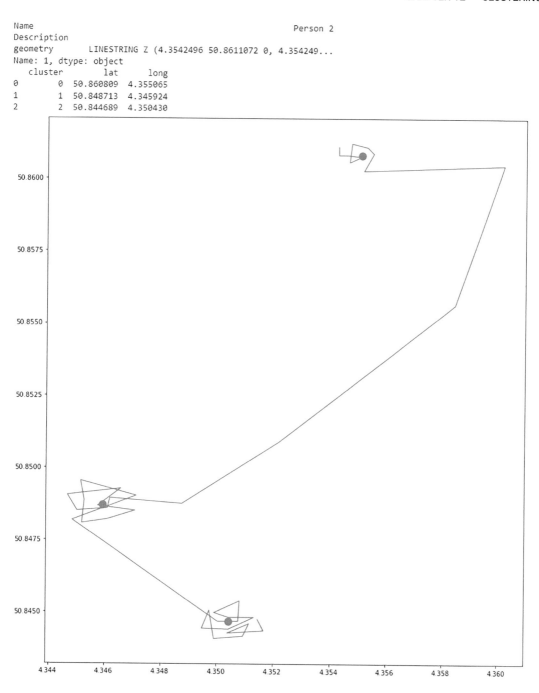

Figure 12-15. *The three central points of Person 2 against their trajectory. Image by author*

For Person 2, we can see that there are three points of interest, and the OPTICS model has correctly identified those three centroids. The model is therefore considered successful on this person. Let's now check the output for the third person in Figure 12-16.

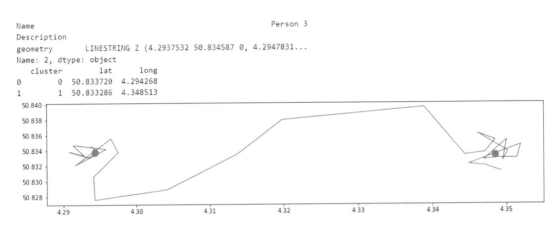

```
Name                                        Person 3
Description
geometry        LINESTRING Z (4.2937532 50.834587 0, 4.2947831...
Name: 2, dtype: object
    cluster        lat        long
0         0   50.833720    4.294268
1         1   50.833286    4.348513
```

Figure 12-16. *The two centroids of Person 3 against their trajectory*

This result for Person 3 is also successful. There were two points of interest in the trajectory of Person 3, and the OPTICS model has correctly identified those two.

Key Takeaways

1. Unsupervised machine learning is a counterpart to supervised machine learning. In supervised machine learning, there is a ground truth with a target variable. In unsupervised machine learning, there is no target variable.

2. Feature reduction is a family of methods in unsupervised machine learning, in which the goal is to redefine variables. It is not very different to apply feature reduction in spatial use cases.

3. Clustering is a family of methods in unsupervised machine learning that focuses on finding groups of observations that are fairly similar. When working with spatial data, there are some specifics to take into account when clustering.

4. The OPTICS clustering model with haversine distance was used to identify points of interest in the trajectories of three people in Brussels. Although the default OPTICS model did not find those points of interests correctly, a manual tuning has resulted in a model that correctly identifies the points of interest of each of the three people observed in the data.

CHAPTER 13

Conclusion

Throughout the 12 chapters of this book, you have been thoroughly introduced to three main themes. The book started with an introduction to spatial data in general, spatial data tools, and specific knowledge needed to work efficiently with spatial data.

After that, a number of common tools from Geographic Information Systems (GIS) and the general domain of spatial analysis were presented. The final chapters of this book were dedicated to machine learning on spatial data. The focus there was on those decisions and considerations in machine learning that are different when working on machine learning with spatial data.

In this chapter, we will do a recap of the main learnings of each of the chapters. At the end, we will come back to some next steps for continuing learning in the domain of machine learning, data science, and spatial data.

What You Should Remember from This Book

In this section, we will quickly go over the main notions that have been presented throughout the book. As some topics were presented only in a specific chapter and did not come back multiple times, this will help you refresh your mind and give you pointers to where you can find some of the key information in case you need to go back.

Recap of Chapter 1 – Introduction to Geodata

Chapter 1 started with a general introduction to geodata, with a number of definitions. Many of those definitions will be useful in your daily work with spatial data, including

- Cartesian coordinates

- Polar coordinates and degrees

- Euclidean distance and haversine distance

J. Korstanje, *Machine Learning on Geographical Data Using Python*,
https://doi.org/10.1007/978-1-4842-8287-8_13

The chapter then moved on to introduce a number of standard tools for GIS analysis including ArcGIS, QGIS, and other open source software. Python was used as a tool in this book, and there are numerous convincing reasons to use it. If you want to become a GIS analysis expert, it may be useful to learn other tools as well, but we will come back to options for future learning paths later in this chapter.

Chapter 1 moved on to introduce multiple data storage types for spatial data:

- Shapefiles

- KML files

- GeoJSON

- Image formats (TIFF/JPEG/PNG)

- Standard formats (CSV/TXT/Excel)

When working with Python, we generally have much more freedom of data types, as we are able to program any processing operation that we could need. This is not always the case in click-button environments. Anyway, it is important to be able to interoperate with any data storage format that you may encounter.

The chapter concluded by presenting a number of important Python libraries for working with spatial data in Python, some of which were used extensively throughout the other chapters of the book.

Recap of Chapter 2 – Coordinate Systems and Projections

In Chapter 2, we went into more detail on coordinate systems and projections. Spatial references in data are what differentiate spatial data from nonspatial data. Coordinate systems are one of the things that make working with spatial data difficult. Even if you have a latitude and longitude column in your dataset, there are many ways to interpret this, depending on coordinate systems.

When making maps, you are also encountering the theoretical problem of showing a part of the globe that is essentially a 3D object onto a 2D image. This is impossible, and we need to choose among multiple projections that are each wrong in their own way. In general, maps try to respect one criterion among the correct area, correct distance, correct shape, and correct direction. We need to find one that best corresponds to our need. The coordinate and projection systems that were covered are

- WGS 1984 Geographic Coordinate System

- ETRS89 Geographic Coordinate System

- Mollweide equal area projection (a.k.a. Babinet)

- Albers equal area conic projection

- Mercator conformal projection

- Lambert conformal conic projection

- Azimuthal equidistant projection

- Equidistant conic projection

- Lambert equal area azimuthal

- Two-point equidistant projection

There are many coordinate systems out there, and finding the one that corresponds best to your need can be a challenge. In case of doubt, it may be best to stick to the more common projections rather than the more advanced, as your end users may be surprised if the map doesn't correspond to something they are familiar with.

Recap of Chapter 3 – Geodata Data Types

In Chapter 3, you have learned about the different data types that geodata can have. A first important distinction is between vector data and raster data. Raster data is image-like data in which you divide the surface of your map in pixels and assign a value to each pixel. This is often used for scientific calculations like hydrology, earth sciences, and the like. A heat map is a great example of this, as a heat map does not have a specific shape. Rather, you want to specify heat at each location of the map.

Vector data works very differently. In vector data, you have objects that are georeferenced. There are three main vector data types: points, lines, and polygons.

- Point data has only one coordinate. A point has no size and no shape, just a location.

- A line is a sequence of points. A line consists of straight lines from each point to the next, but the overall line can have a shape that consists of many straight-line parts. The line also has a length, but it has no surface. The width is zero.

– Polygons are any other form, as they capture a surface in between maps. The polygon consists of a sequence of points that together make up the border of the polygon. All area inside this is part of the polygon. Polygons have a border length and a surface area.

Choosing the data type for your project will really depend on your use case. Although conversions between data types may sometimes be not super easy to define, necessary cases for conversion happen, and you can generally use some of the GIS spatial tools for this.

Recap of Chapter 4 – Creating Maps

Chapter 4 proposed a deep dive into making maps using Python. As Python is an open source language, there are many contributors maintaining libraries for making maps. In the end, you can choose the library of your choice, but a number of approaches have been covered.

The first approach was to use geopandas together with matplotlib. Geopandas is a great Python package for working with geodata, as it closely resembles the famous Pandas libraries, which is a data processing library that is very widely used in data science. Matplotlib is a plotting library that is also widely used in the data science community. The combination of those two libraries is therefore greatly appreciated for those with experience in the data science ecosystem.

Secondly, you have seen how to build maps with the Cartopy library. Although it may be a bit less intuitive for data scientists, it still proposes a lot of options that are more anchored in the field of GIS and spatial analysis. This approach may be a great choice for you if you come from a spatial analysis or cartography background.

Plotly was proposed as a third option for mapping with Python. Plotly is purely a data visualization library, and it approaches mapmaking as "just another visualization." For users that put strong importance to the visual aspect of their maps, Plotly would be a great tool to incorporate for making maps.

As a fourth mapping tool, we looked into Folium. Folium is great for making interactive maps. When you create maps with Folium, you create a visualization that will almost have a Google Maps–like feeling to it. Of course, this will not be usable in reports or PowerPoints and has a specific use case in use cases that are delivered digitally. It is great for data exploration as well.

Recap of Chapter 5 – Clipping and Intersecting

In Chapters 5 to 8, a number of GIS spatial operations were presented. Chapter 5 started this with the clipping operation and the intersecting operation.

The clipping operation allows you to take a spatial dataset and cut out all parts of the data that are outside of boundaries that you specify to the clip. This even works with features (lines, polygons) that are partly inside and partly outside of your boundaries, and it will make an alteration to those shapes.

Intersections are a spatial overlay operation that allow you to take two input datasets and only keep those features in the first dataset that intersect with features in the second dataset. This operation, just like many other spatial overlay operations, is derived from set theory.

In the Python examples of this chapter, you have first seen how to apply a clip to a dataset with a number of spatial features. In the intersecting example, you have seen how to find the intersections of a river with a road in order to find out where to find bridges. When a road and a river intersect, we must have a bridge or a tunnel.

Recap of Chapter 6 – Buffers

Chapter 6 proposed the buffering operation. Buffers are areas around any existing spatial shape (point, line, polygon). Creating buffers are commonly used in spatial operations as they allow you to create a shape that contains not just the original feature but also its close vicinity.

In the Python use case of this chapter, you have used multiple datasets to define a region in which you want to buy a house based on distance criteria to parks, metro stations, and more. You have used an intersection here to find a location that respects all of the criteria at the same time.

Recap of Chapter 7 – Merge and Dissolve

Chapter 7 covered the merge operation and the dissolve operation. The chapter started with the different definitions of merging. The spatial join is more than just putting different features in the same dataset, as it will also combine the fields (columns) of different datasets together based on spatial overlap. This is a very useful operation, as we are used to be able to join tables only when we have a common identifier between the two tables. For spatial data, this is not necessary as we can use the coordinates to identify whether (parts of) the features are overlapping.

The dissolve tool is different as it is not meant to do join-like merges. Rather, it is useful when you want to combine many features into a smaller number of features, based on one specific value. It is like a spatial group by operation.

Recap of Chapter 8 – Erase

The fourth and last chapter on spatial operations was Chapter 8, presenting the erase operation. Although different definitions of erasing exist, we covered a spatial erasing operation in which you want to erase a specific part of a spatially referenced dataset. This is done not based on a common identifier, nor by removing an entire feature (row) at once, but rather by altering the data to keep those parts of features that must be erased and keep those parts of a feature that must be kept.

Recap of Chapter 9 – Interpolation

From Chapter 9, we moved on to more mathematical topics, even though interpolation is a very common use case in spatial operations as well. Interpolating is the task of filling in values in regions where you have no measurements, even though you have measurements in the vicinity.

Interpolation is a widely studied topic in mathematics, and one can make interpolations as difficult as one wants. The different methods for interpolation that were covered are

- Linear interpolation

- Polynomial interpolation

- Piecewise polynomial interpolation, a.k.a. spline interpolation

- Nearest neighbor interpolation

- Linear Ordinary Kriging

- Gaussian Ordinary Kriging

- Exponential Ordinary Kriging

In the Python example of this chapter, we looked at a benchmark that compares multiple of those methods and see how their results differ. As we have no ground truth in interpolation, one needs to find a way to determine which method is best to use.

Recap of Chapter 10 – Classification

In Chapters 10 and 11, we covered the two main machine learning methods of the family of supervised machine learning. Supervised machine learning is a branch of machine learning in which we aim to predict a target value by fitting a model on historical data of this target variable, as well as a number of predictor variables.

In classification, we do so with a target variable that is categorical. We have covered specific models and metrics for the case of classification.

The Python example on classification covered a use case in which we used movements of clients of a mall, and we used this to fit a predictive model that was meant to predict whether a person would be interested in a product based on their movement patterns. This example used spatial operations from earlier chapters to do the data preprocessing for the model.

Recap of Chapter 11 – Regression

Regression is the supervised machine learning counterpart of classification. Whereas the target variable is categorical in classification, it is numeric in regression. You have seen numerous metrics and machine learning models that can be used for the regression case.

The Python example of this chapter used spatial and nonspatial data to predict Airbnb prices of apartments in Amsterdam. You have seen an example here of how to prepare the spatial and nonspatial data to end up with a dataset that is suitable for building regression models. You have also seen how to fit, evaluate, and benchmark multiple machine learning models for regression.

Recap of Chapter 12 – Clustering

In the last chapter, you have seen a third machine learning method called clustering. Clustering is quite different from regression and classification, as there is no target variable in clustering. As the method is part of the family of unsupervised models, the goal is not to predict something, but rather to identify groups of similar observations based on distances and similarities.

Clustering on spatial data comes with some specificities, as computing distances on spatial data (coordinates) needs to be mathematically correct. As you have seen in the earlier chapters of the book, distance between two coordinates cannot be correctly computed using the standard, Euclidean distance.

The Python use case in this chapter presented how to use the OPTICS clustering model with the haversine distance metric to create a clustering method that needs to find points of interests in movement patterns of three GPS-tracked people. This chapter has concluded the part on machine learning on spatial data.

Further Learning Path

Of course, this book should have given you a solid introduction into spatial data with Python and machine learning on spatial data, but there is always more to learn. This section will give you some ideas for further learning. Of course, there is never one only way for learning, so you may consider this as inspiration rather than a presentation of the one and only way to proceed.

As the book has touched mainly on spatial data and machine learning, two interesting paths for further learning could be specializing in GIS generally or going further into machine learning. Otherwise, things like data storage and data engineering for spatial data can also be interesting, although a bit further away from the contents of this book.

Going into Specialized GIS

If you want to go on to become specialized in GIS and spatial analysis in general, I suggest learning specific tools like ArcGIS and QGIS, or at least start by checking out those tools and getting around the basics of them.

In this path, it is essential to spend time learning spatial statistical methods including spatial autocorrelation, spatial heterogeneity, kriging and other interpolation tools, and more theory to spatial analysis.

It may also be worth it to spend more time learning about mapmaking, as the resulting deliverables of this domain will very often be maps. Mapmaking can be easy, but great maps take time and effort, as they are ultimately tools for interpretation and communication. Your time will be well spent mastering this skill.

Specializing in Machine Learning

Although this book has touched multiple machine learning methods applied to spatial data, the topic of machine learning is a very complex mathematical domain. If you want to become an expert in machine learning, there are many things to learn.

This career path is quite popular at the moment, as it allows many people to work on the forefront of present-day technology, yet it must be noted that there is a serious learning curve for getting up to speed with this domain.

Remote Sensing and Image Treatment

Remote sensing was not covered in this book, but thanks to recent advances in computer vision, there are a lot of advances in earth observation and remote sensing as well. As you probably know, a large number of satellites are continuously orbiting around the earth and sending photos back. With new technologies from artificial intelligence, these photos can be interpreted by the computer and can serve a large number of purposes.

Although computer vision may seem like a scary and difficult domain, you can see it as a next step in your learning path after mastering "regular" machine learning. Of course, the learning curve may be steep here as well, but the use cases are often very interesting and state of the art.

Other Specialties

There are also other fields of study that are related to the topics that have been touched on in this book. As an example, we have talked extensively about different data storage types, but we have not had the room for talking about things like specific GIS databases and long-term storage. If you are interested in data engineering, or databases, there is more to learn on specific data storage for spatial data, together with everything that goes with it (data architectures, security, accessibility, etc.).

Many other domains also have GIS-intensive workloads. Consider the fields of meteorology, hydrology, and some domains of ecology and earth sciences in which many professionals are GIS experts just because of the heavy impact of spatial data in those fields.

When mastering spatial data operations, you will be surprised of how many fields can actually benefit from spatial operations. Some domains already know it and are very GIS heavy in their daily work, and in other domains, everything is yet to be invented.

Key Takeaways

1. Throughout this book, you have seen three main topics:

    a. Spatial data, its theoretical specificities, and managing spatial data in Python

    b. GIS spatial operations in Python

    c. Machine learning on spatial data and specific considerations to adapt regular machine learning to the case of spatial data

2. This chapter has presented a number of ideas for further learning in the form of potential learning paths:

    a. Specialize in GIS by going into more detail of different GIS tools and mapmaking.

    b. Specialize in machine learning by studying machine learning theory and practice in more detail.

    c. Going into advanced earth observation use cases and combining this with the study of the field of computer vision.

    d. Other ideas include data engineering by focusing on long-term efficiently storing geodata or any other field that has a heavy component of spatial data, like meteorology, hydrology, and much more.

Index

A

Albers equal area conic projection, 33, 34, 299
Azimuthal equidistant projection, 36–37
Azimuthal/true direction projection, 38–41

B

Babinet projection, 33
Buffering operations
  data type, 128
  definition, 128
  difference creation, 147
  GIS spatial operations, 301
  intersection operation, 129
  line data, 130, 131
  point data, 129, 130
  polygon, 132
  Python
    data resulting, 144
    house searching criteria, 132
    LineString object, 137–141
    point data, 133–137
    polygons, 136, 141–146
    visualization, 143
  schematic diagram, 128, 129
  set operations, 146–149
  standard operations, 127

C

Cartesian coordinate system, 5, 6
Cartopy, 88–92
Classification, 225
  data modeling
    array resulting, 243, 244
    dataframe format, 242
    error analysis, 247
    logistic regression, 245
    plot resulting, 247
    predictions, 245
    resulting comparison, 246
    stratification, 244
  GIS spatial operations, 303
  machine learning, 225
  model benchmarking, 248–250
  reorganization/ standardization, 238–242
  spatial communication
    advantage/disadvantage, 233
    data resulting, 235, 236
    feature engineering, 227
    geodataframe, 235
    importing data, 227–232
    map resulting, 230
    operation, 232, 233
    resulting dataframe, 238
    source code, 237
    truncated version, 232
    use case, 226
    wide/long data format, 233

© Joos Korstanje 2022
J. Korstanje, *Machine Learning on Geographical Data Using Python*,
https://doi.org/10.1007/978-1-4842-8287-8

D

Printed in the United States
by Baker & Taylor Publisher Services